No Time
and Nowhere

No Time
and Nowhere

Fergus Hinds

BOOKS

Winchester, UK
Washington, USA

First published by iff Books, 2016
iff Books is an imprint of John Hunt Publishing Ltd., Laurel House, Station Approach,
Alresford, Hants, SO24 9JH, UK
office1@jhpbooks.net
www.johnhuntpublishing.com
www.iff-books.com

For distributor details and how to order please visit the 'Ordering' section on our website.

Text copyright: Fergus Hinds 2015

ISBN: 978 1 78535 185 3
Library of Congress Control Number: 2015953281

A CIP catalogue record for this book is available from the British Library.

Design: Stuart Davies

Printed in the USA by Edwards Brothers Malloy

We operate a distinctive and ethical publishing philosophy in all
areas of our business, from our global network of authors to
production and worldwide distribution.

CONTENTS

Other Works

RICHES FROM WRECKS
The Recovery of Sunken Cargoes

I am pleased to acknowledge that this book contains a number of cases taken from *Apparitions* by Celia Green and Charles McCreery, published by the Institute of Psychophysical Research.

The IPR is an independent academic research organisation, based in Cuddesdon near Oxford, which aims to study the psychology and physiology of quasi-perceptual experiences, and which depends on private rather than governmental funding.

Potential supporters or voluntary workers who would like more information about the IPR should visit:

www.celiagreen.com

Preface

The Question

An immaterial domain behind this one is nothing new. That idea has underpinned all religions since human beings first began – and its latest formulation is the parallel universe. Physical evidence shedding new light on this domain is the suggestion of two kinds of measureable physical activity that have no physical causes.

One is brain activity detectable on an electro-encephalogram which relates to thoughts that just appear, they are neither self-generated nor induced by sensory inputs from outside our heads. Where such thoughts relate to contemporary affairs the non-physicality manifests itself as the unaccountable conveyance and sometimes storage of information; where they concern future events the question is the apparent conveyance of information through time. The second unaccountable physical activity is the peculiar behaviour of matter on a very small scale in physics laboratories, its apparent non-physicality being observable as derangements of space or time – once again mediated by information. The two are not at all the same thing but their shared physical acausality is curious.

* * *

J.W. Dunne was an aircraft designer in the early years of the twentieth century, a practical man not in the least credulous or fey. In 1927 he published a remarkable book called *An Experiment with Time* which has been in print continuously ever since. It arose from a few striking dreams he had which seemed to presage later events, and then his discovering that other people had them too.

If precognitions stand scrutiny there is its startling temporal anomaly to contend with, but that masks another less dramatic but equally inexplicable feature of it – the conveyance of information in an unaccountable way. Such information has three distinct presentations in waking psychological contexts. One of them is shared hallucinations.

To understand the enormity of these experiences imagine a solitary healthy man looking up from a football report in his newspaper to see the figure of a muddy and dispirited soldier in battledress standing in the corner of his room when there is no man there and no man-like object he could mistake for one. He is hallucinating, which is to say he is creating the image is his own head without any ocular input to the process. Now imagine three people in company all seeing without prompting the same non-existent figure in the same part of their environment for the same length time. That has happened occasionally, and when it does it poses this question: 'Why does each person create for himself the same thing?' What is the information specifying the image and how does it get into the heads of these observers?

The same questions on conveyance and uniformity can be asked of 'ghosts' – confining that term to the same figure repeatedly appearing in the same place to different people on different occasions, sometimes to people who knew nothing of it before. Once again, why that particular image and not some other? Where does the specification lie between appearances? How does it find its way into their heads? First it is not there and then it is – something is arriving to account for the uniform imagery.

The detail in all these experiences is such that the uniformity cannot be chance, and there are no analogies to allow even an ordinary guess at an explanation. Nor can these visual experiences be dismissed out of hand, for realities experiences through our senses are not the only realities. Were the headless wraith of Charles I seen drifting down Whitehall some January night not much could be said or inferred about the spectre itself, but the image of it is a fact in the observer's mind and grist to the mill of contemporary investigation. The essence of all ghosts is images and not external objects, though they are none the less real for that. Visual illusions misconstruing real ocular inputs are quite different to these that have none; spontaneous pathological hallucinations repeat and are distressing; the cause of those induced by ingesting hallucinogens is conspicuously physical. Each can be easily differentiated from these unprompted one-off experiences of healthy individuals.

The specification for the collective and successive images is not conveyed by sight, neither grossly nor subliminally, or everyone would see them and there would be photographs of them. It is not by ordinary speech when there is more than one observer present because it takes more words to describe scenes than the duration of these experiences generally allows time for. It is not by sound directly because we do not image sound visually. It cannot be by touch, smell or taste for the same reason and because they are confined to contact sensations. Thus the specification is not being conveyed in any sensory way, and it takes only a moment to run through all the other recognized physical influences to see that it is not transmitted by any of them either. We understand our physiology well enough to know that we have no receptors that allow us to enjoy any *adventitious* images – as distinct from activating ones already there – that are conveyed by compression waves other than sound, or by electro-magnetic radiation outside the visual frequencies, or electric currents, electrical or magnetic fields, or cerebral probes. It is not by radiated particles, nuclear forces or gravity either. There is nothing else; for this externally sourced information non-sensory implies non-physical.

There are only three ways to account for the reports of these hallucinations. The first and commonest is to suppose that the figures do not exist at all, and therefore that the observers did not experience what they thought they did and were suggestible or credulous and mistaken. There is no evidence for that assumption, which is really a reluctance to hold any view at all. The next supposition is that all apparitions have a physical presence; it is true that the observers report external realities with at least enough physicality to reflect light, but that is only their inference and as a hypothesis it flatly contradicts what we know very well of the physical universe. It has been current for centuries but has led nowhere in all that time. The phenomenon is not an external figure but an image internal to that mind only, and as such not easily refuted. The third option is to see shared hallucinations as psychological self-creations harmonized between individuals by detailed information conveyed in some unrecognized way. To make any progress there is nothing for it but to countenance that idea as a possibility – because there is no other – and see where it leads. The ground is unfamiliar but it is the only path untrodden.

Commoner than hallucinations or ghosts are so called 'crisis cases' – those intimations of distress to a distant friend or relative which arrives at the moment he is afflicted. How is the information transmitted, why to only one person, and why only occasionally? There are simply no answers. These experiences seem to be common but so little information is conveyed – seldom more than unspecified distress to that person at that moment – that there are no collections of records dedicated to them.

And then there is apparent precognition, of which there are more than five thousand English language records in the twentieth century, about two thirds of them from the United States. The non-physicality of the information in all these experiences allows them to be perceived as distinct classes or categories of phenomena having the concomitant of detectable *physical activity* in a human brain, subject always to the observations being reliable.

Are they reliable? The spontaneous psychological experiences of these kinds never fail to be considered until after they have happened. Nothing about them can be recorded or measured, as opposed to recalled, and since measurement is central to all the hard sciences, no measurement means no science. The phenomena are historical events and as such can never be proved logically like Pythagoras' theorem, or by repeatable experiment like Ohm's law. History is not accessible to the established procedures of science, but that does not mean we cannot deal with it at all. Every day law courts get by perfectly well with an entirely different approach; they do not demand that a murderer repeat his crime before they can convict him of it. If one cannot countenance some alternative to the strictest scientific rigour then the working of these experiences is forever beyond the reach of human minds.

That cannot be the case, whence a different approach is needed.

There are only two main questions. First: what can be done to verify the receipt of contemporary information in inexplicable ways? Second and quite separately, is information that apparently concerns future events really doing so, bearing in mind if it were our understanding of the universe would have to accommodate the implied derangement of time?

Part One of this book presents and analyses various examples of contemporary psychological waking experiences, and then addresses an apparently non-physical effect evident only on a very small scale in physical laboratories.

Part Two presents reports of precognition awake and asleep, including an outline of current dream research to distinguish ordinary features of that medium from anything else that may not be ordinary, and concludes that the possibility of a temporal anomaly in psychological contexts has to be faced; and there is a suggestion of temporal derangement from those same laboratories also.

Part Three develops a hypothesis that there are signs of a distinct immaterial, timeless and locationless world behind or beside the physical one. Even if such a world were to be proved it would not establish the nature of non-physical information, but it would provide a feasible venue for it that eliminates the need for unexplained transit through space or backtracking through time. It would permit knowledge of events at a distance. It could obviate the implied paradox of altering the future that has been disclosed and restore free will; it could embrace the unaccountable physical behaviour of matter on a small scale; it might also suggest new ideas in other fields as well. It is a starting point for quite new enquiries.

* * *

In the course of examining the effects of static charges on animal tissues in 1780, the Italian physicist and physician Galvani noticed that the legs of his recently dissected frogs would twitch occasionally. Though curious he did no more than establish that movement appeared only in fresh specimens touched simultaneously with two dissimilar metals, but he mentioned it to his friend Volta who was professor of physics at Pavia University. In time Volta discovered that the animal tissue mattered only for its moisture being an electrolyte, and by 1794 he had developed an electric cell incorporating a much better one. Behind those little spasms lay the unimaginable phenomenon that underlies today's entire technical civilisation – flowing electricity. Things that do not fit the pattern must be interesting; anything really new is necessarily unheralded and unforeseeable, and any first approach to it initially tentative.

* * *

Many of Dunne's experiences feature in this book because it was his ideas that prompted it. I first read *An Experiment with Time* in my late teens. About twenty years' later I was wondering what developments there had been in the light of new discoveries and decided that anything substantial would have come to my notice somehow; and a year or two later still I decided that I would take a look myself to see whether that remained the case.

My working life consisted of ten years as a hydrographic surveyor in the Royal Navy followed, after a couple of years in banking, by a very specialized kind of marine salvage. I come to this topic with an entirely open mind – or no particular expertise at all, as you will. I am not religious, superstitious or gullible.

* * *

I owe thanks to Sally Feather for permission to use material collected by her mother Louisa Rhine, the greater part of it from *The Invisible Picture*; and to Dr Keith Hearne for permission to use some of his examples from his *Visions of the Future* to illustrate parts of my argument. I make brief mention of cases collected by the playwright J.B. Priestley and by Dr Rupert Sheldrake, my thanks to them and to the American Journal of Physics for permission to reproduce the illustration from Tonamura's experiment.

Many thanks also to, my wife Amanda for her comments and editorial help.

Introduction

Hallucinations and ghosts and those experiences that used to be called telepathy have all been open to observation for centuries. Why have they not been resolved before now?

To put hallucinations in their historical intellectual context one could say that the shift of emphasis towards western science was largely accomplished between perhaps 1850 and 1925. There had been significant discoveries before that – Newton's mechanics, some chemistry, electro-magnetic induction, the foundations of geology among them – but at the beginning of that period questions such as: Why wolves? Why gales? Why infections? Why a blue sky? All had one and the same answer – the unfathomable will of God. But by the end of it, they and the answers to thousands like them were each in crisp and rational focus. At the start of that remarkable age it must have seemed that any well-ordered set of observations would soon be scientifically explained. Not a few in those days saw those things which now seem outside the purlieus of science as no more extraordinary than many natural phenomena. There could at first have been no very good reason to categorize things like ghosts much differently.

Investigators faced two difficulties. One was the widespread belief in an afterlife that beguiled many into supposing that so supremely important an idea might be proved scientifically. Necromancy (calling up the dead) seemed a promising line to pursue, especially since professional mediums claimed that it was generally repeatable to order. In the late nineteenth century mediums were in great demand. It seems extraordinary to us that these tricksters should have been able to get away with their simple deceits, but there was a credulous craze for them, much as for unidentified flying objects (UFOs) after World War II. Mediums were illusionists producing in the dark sounds and movements attributed by them to departed spirits. They did it for entertainment and money, and some of them did very well. Professor Hansel (Hansel 1989) says that in the United States in 1900 mediums were the second highest-paid self-employment for women; the Italian Eusapia, perhaps the most famous of all, commanded $125 per hour in 1910, the equivalent of £7000 or so nowadays. Accounts of how she and other famous ones were exposed as frauds make entertaining reading. Many prominent scientists were deceived; professional illusionists – among them Houdini – were not.

However laughable it looks now, the idea that a physical explanation of 'paranormality' was just around the corner continued to be perfectly reasonable until science collectively had acquired a comprehensive understanding of the natural world without finding one. For a long time it was obviously impossible to know whether that moment had arrived. By the time most reckoned it had – 1920 to 1930 say – credulity and associations with other nonsenses like astrology, not to mention a smattering of fraud, had brought the whole field into disrepute.

The other difficulty with this class of problems was the general proposition that the existence of a real effect had to be proved before its mechanisms could be investigated. It was supposed that demonstrating information being conveyed between individuals in experimental conditions that precluded transfers by recognized means would constitute that proof. Trials at first involved blindfolds, ear-plugs, wrapping in blankets, and so on. The supposedly gifted subjects were paid a wage or fee as long as the effects lasted and since they were allowed to object to any constraint that 'disrupted their powers', most of them found they could oblige.

Attempts to influence dreams that were first made in the United States in the 1950s, and more recently in Ganzfeld experiments whose essential characteristic is a degree of sensory deprivation for the recipients, were beset with difficulties in assessing how close the transmission and the receipt really were. If a line drawing of the setting sun just on the horizon were received as what looked like the side view of a two-wheeled cart on a long level road, was that interestingly close or an absolute miss? There has been work with animals as well.

None of these approaches has worked to any noteworthy extent. As with necromancy a few of the proceedings degenerated into conjuring, a consequence only of failure.

A variation on all of this was confined to telepathy and known as extra-sensory perception – ESP. It was supposedly repeatable and quantifiable through the use of Xener cards, which come in twenty-five card packs, each showing one of five simple symbols designed to be emotionally neutral. Subjects were invited to communicate to someone else what shape they were viewing. The outcome of thousands of such tests was, in some 'gifted' individuals, a success rate two or three per cent above chance levels, with a repeatable decline or fade in prolonged sessions. So small a margin can be called in question by the least experimental imperfections, and even if there is any real effect it is being masked by huge irrelevancies and has no practical use.

Conspicuously absent from these enquiries have been spontaneous transfers of information rather than those that were deliberately prompted. Apparitions have received little attention beyond collecting instances of them. There has been no categorising of their different presentations and no detailed reflection upon observational imperfections in any instance; and so also with what are called 'crisis cases' – intimations of distress or injury to a distant relative or friend.

Precognition, succinctly the apparent backtracking of information through time, has excited attention throughout the ages. Aristotle was the first to write about it; he held that it could only concern human actions; and it did not come from God because those who had them were not the 'best and wisest, but merely commonplace persons'; and that foreknowledge of events beyond human agency – something like an earthquake– were all coincidences. Dunne's reasoning led him to a temporal series – Times, 1, 2, 3, etc. – but he could not escape the paradox highlighted by Newton, that if time flows there must be another time to measure that flow – and so *ad infinitum*. Jon Taylor (Taylor, 2007) has related precognition to Blom's 'block universe', but he was obliged to say that there can be no foreknowledge of an event that the subject tries or intends to prevent. But sometimes there is.

Of all these psychological experiences that fall to be investigated, crisis cases are the commonest of all but because very little information is ever conveyed in them their study does not promise much by way of a dividend. Visual hallucinations are free of that objection because they require, in pixel terms, a great deal of information. Of them successive hallucinations – ghosts – have the complication of association with a place and seemingly also with the past, and premonitions are complicated not only by the implicit temporal anomaly but also (most often) for their dreamed presentation. Waking visual hallucinations common to several individuals are the best introduction to non-sensory information overall. There follows a full report of one from each main class with briefer reports to illustrate particular points. They and the laboratory curiosities lead into quite unexpected territory.

References and Citations

The Works Cited in the text are listed by author and date. Each unusual psychological experience is marked in the text by the symbol 'ψ' and its source listed numerically.

Part I

Contemporary Experiences

Analysis in detail of one collective, successive, and crisis intimation, each followed by two or three other instances in brief; then minor classes of hallucinations and the mechanisms of them all, followed by an instance of inexplicable behaviour in physical laboratories.

Chapter One

A Collective Hallucination and notes on three others

This is the most straightforward example of co-ordinated non-ocular imaging while awake that I have come across. In the course of a radio broadcast on apparitions and hauntings in 1936 Sir Ernest Bennett asked listeners to send him accounts of any experiences they had had. A Miss Anna Godley who lived in Leitrim in Ireland supplied this one. Her written report comes first, and then those of her two companions.

(Robert Bowes ψ1) One afternoon in February 1926 I went to visit a former old farm labourer of mine, Robert Bowes, who lived about a mile away but inside the place; it was about 2.30. He had been ill for some time but was not any worse. I had lately broken my leg and was in a donkey trap, the steward was leading the donkey and my masseuse walking behind. I talked to Robert through the open window and he sat up and talked quite well, and asked me to send for the doctor as he had not seen him for some time. I then came straight back. The road runs along the shores of a big lake and while the steward stopped to open a gate there, he asked me 'if I saw the man on the lake'. I looked and saw an old man with a long white beard which floated in the wind, crossing to the other side of the lake. He appeared to be moving his arms, as though working a punt, he was standing up and gliding across but I saw no boat. I said 'Where's the boat?' The steward replied 'There is no boat'. I said 'What nonsense, there must be a boat, and he is standing up in it,' but there was no boat he was just gliding along on the dark water; the masseuse also saw him. The steward asked me who I thought he was like, I said 'he is

exactly like Robert Bowes, the old man.' The figure crossed the lake and disappeared among the reeds and trees on the far side, and we came home. I at once went to take off my hat and coat and to write a note for the doctor, but, before I left my room the bell rang and the doctor came in. I said I was glad to see him as I wanted him to go and see Robert Bowes; he said 'I have just been there' (he went in a car by a different road to the one I had been on) 'and the old chap is dead.'

That is all; but no living man crossed the lake, and there was no boat on it… He (Robert) was quite plainly seen by all three of us. He was a remarkable-looking, handsome old man, full of Irish wit and humour and I must have seen him just as he was leaving this world. He has never been seen again.

The steward Robert Gallagher wrote on 3rd January, 1937:

…visiting a sick man on the estate. We had no idea that he was so near passing out of this earthly life, but on parting from him on his sick bed, and on our way home we were amazed when passing Killegar Lake, at the close of the same day, to see him walking on the surface of the water. His whiskers were floating in the breeze and when near the shore, in the shadow of the wood, he completely disappeared. We all three beheld the same sight.
(Author's note; sunset on that day was at about 1635.)

And on the same day the masseuse Miss Goldsmith wrote:

We had just left the cottage where the old man was lying in a terribly weak condition, and on walking back, we were impelled to glance towards the lake, and saw a shadowy, bent form step from the rushes, and into a boat, and after an interval, just disappear. We learnt later on that the old man had passed away at that moment. Though not in the least

15

given to seeing visions, but being of an extremely practical nature, I certainly saw the spectre as I describe it.

Miss Godley wrote again on 11th March, 1938 to clarify a couple of points:

1. There was no boat on the lake when we saw Robert Bowes crossing. My masseuse thought he was standing up in one, but he wasn't; there was nothing except what looked like a pole being used by him to help himself across.
2. Robert Bowes had a white beard; in this country the people always call a beard 'whiskers'. Robert had both, and his beard streamed in the wind. Both beard and whiskers were white. There was no boat on the shore after we had seen the figure on the water.

In 2000, I telephoned Lord Kilbracken who was a cousin of Miss Godley and had inherited the estate on which she had lived. The Bowes episode was well known in the family. The big house looked out on the lake, the cottage was still standing as a ruin, and the gate was still there too. Lord Kilbracken said the figure must have been less than one hundred yards distant.

On the face of it the three of them were induced to create for themselves in the same general location and for the same length of time similar images of Robert Bowes crossing the lake when there was no one there, but that experience is a past event that cannot be proved by reasoned argument nor the facts confirmed by experiment. Our only option is to use personal judgement to balance probabilities.

The preliminary difficulty with all hallucinations is determining whether they are anything but observational aberrations. A real figure on the lake exactly as perceived, with a light-reflecting surface, with density and therefore with mass, is a supposition that has had its day. Physics does not allow it, it is impossible, and in any case a material entity would be much more difficult to explain than its psychological counterpart. The orthodox alternative is mistaken observation. If that is what happened a psychologist would expect to find among this muster of all the ordinary possibilities that follow one that would explain this and every other similar case.

1. The uniformity of the hallucination between observers can be attributed to chance.

2. The image could have been an illusion, a real external but indistinct object being misconstrued due to poor light, mist, rain, fogged spectacles, even bad eyesight, any of them perhaps coupled sometimes with exhaustion, fear, or anxiety; the image would be internally created and correctly reported but a mistaken interpretation of the inputs that had triggered it. As Shakespeare puts it in *A Midsummer Night's Dream* 'how easy is a bush supposed a bear'. The spurious image has to be close to a real one and not too implausible – a bush at night may be supposed a bear but not a cottage or a waterfall; nor may a static bush become a moving bear. If the illusory image moves any distance its physical trigger must move too. And if the experience is to be accepted afterwards there must also be some interruption to prevent the mistaken image being resolved in an ordinary way, such as the observer approaching the object more closely or studying it more attentively.

3. There was no external object but the observer imaged his or her expectations without any ocular input.

4. The reported image never existed at all. There are four possibilities:
   (a) The observer was lying.
   (b) Unprompted spontaneous self-deception, perhaps founded on generalized superstitions or beliefs

unrelated to any specific expectations in the actual circumstances, might sometimes be sufficient cause.

(c) The reported image was not seen but the substance of it was induced unintentionally in the minds of suggestible observers by other people's comments at the time or later.

(d) A different image was observed but significantly misremembered later, or hugely modified by repeated discussion.

5. The cause of the images was pathological – mental derangement, alcohol, or drugs.

This is a review of the trio's experience against that list:

Chance

The lady of the estate writes as though this was her first experience of a hallucination and the masseuse specifically says that it was. That these two women should be supposed to have undergone a rare class of experience at the same moment by accident is incredible, as is their being simultaneously subjected to what was for each of them a unique lifetime experience for no reason; nor can I imagine a connection by chance to the third person present. Three people without prompting thinking exactly the same thought at the same time is improbable enough, but that they should all simultaneously image without optical stimulation the same improbable figure strains credulity past its limits.

Expectation

None of the observers could have had any expectation of seeing Bowes walking on water and created an image of him for that reason – this was the twentieth century. Neither the lady nor the steward supposed him near to dying – she saw him 'ill but not worse' and he 'had no idea' that Bowes was nearing his death.

The masseuse saw him as 'terribly weak', but she was dismissive of fanciful imagery. Nor do the two women sound in the least prone to baseless superstitions. On the other hand the possibility of a subliminal indication of death is undeniable. Perhaps Bowes' wish to see a doctor could have suggested it; the image's 'crossing over' – the Styx, the Jordan, to the 'other side' – as a cultural metaphor for dying may suggest that the idea was in their minds already, but were all three toying with the same idea it still gives no reason to create any visual image of anything, much less that particular one simultaneously.

Illusion

Might there have been some real object on the lake each one of the three could each have misconstrued in the same way? As a general proposition there might, but the likelihood of each possibility diminishes on examining its details. If this supposed object were where it appeared to be, its general appearance must have been a man-sized vertical rectangle roughly 1.7 metres by 50 centimetres with a low sub-surface centre of gravity or a substantial submerged volume to keep it upright. There was no gale sufficient to move such a thing at perhaps a metre per second. May be the distance to some much smaller object was over-estimated, and therefore it appeared as larger and faster than it was? Lord Kilbracken put it about one hundred yards away, and the figure stepped into a particular stand of trees at a location that was for two of the observers part of their everyday surroundings. (The steward called them 'the' wood.) It is difficult to countenance errors of scale.

If there was no floating object could the observers have misconstrued something meteorological? Perhaps a patch of mist? That would not have had the necessary shape and density to resemble a man, nor subtended horizontally less than the quarter of a degree which Lord Kilbracken's range requires; and mist would have obscured much else besides. The same goes for a squall of rain which would for practical reasons have been immediately recognized for what it was by a temporally handicapped woman in an open donkey trap. No one reported mist or rain remaining when the hallucination ended.

What about the reflection of a cloud? Or a very localized patch of ripples? Either might have moved or disappeared suddenly, but both are commonplace sights to people who live by water, and in trying to interpret a puzzling image both these possibilities would have occurred to those viewers almost at once in the process of elimination. The same two of the trio who were familiar with that lake had known it for years under all meteorological conditions. And familiarity with the lake generally would allow them to resolve any other uncertain natural sight – as an example a skein of alighting ducks momentarily presenting a narrow vertical outline.

But none of it contributes to Bowes in particular.

And then there is plausibility. The image of Bowes must have endured for fifteen to twenty seconds to allow two of the witnesses in sequence to verify for themselves that they really were seeing so improbable a sight, and then to accommodate the remarks they made about it. A man standing on a lake, in a boat or out of one, is inherently implausible. However indistinct the ocular input, calm, alert, unhurried people mistaking an ordinary sight for an absurdity for that long is improbable to the point of preposterous; and why should the inputs be indistinct on a dry February afternoon about an hour before sunset? For this to be a mistaken visual image of a real object that thing must have been man-sized, vertical, moving, able to vanish without trace, and like enough to Bowes to make each of them create an image of him in particular. There is also the fact that the location of the image seems to have differed between observers. The steward saw a 'man on the lake' who 'near the shore in the shadow of the wood disappeared completely', apparently being on the water when first seen. The lady saw a man 'crossing the lake' to disappear 'among the reeds and trees on the far side' – on his way when first seen because she asks 'where's the boat?' The masseuse saw him 'step from the rushes and into a boat' and 'after an interval just disappear' – starting from the near shore but not apparently completing the crossing. If this was an illusion its physical prompt can only have occupied a single position.

No External Object

(a) If no image of any kind was experienced Miss Godley must have written two pointless lying letters to Sir Ernest Bennett, and then induced her steward and former masseuse to do the same. Why should they? Any one of them, never mind all three. Lord Kilbracken knew his cousin personally and repeated her story without hesitation. They do not sound like liars. Unless there is some category of habitually exuberant raconteurs who join together to spin outrageous yarns, mendacity seems to me straying from the possibilities in this case into new ideas on a different one.

(b) Might it have been a mistaken observation from the start? Not an illusion, a faulty image of some real sight, but the supposition of one based on nothing at all? There have been experiments showing that some witnesses to a contrived commotion grossly mis-report it, even immediately afterwards, creating or adding details from other contexts for which there was no basis in reality. However, they also show three other features. First, the process needs an actual event to initiate it, a real scene is the essence of all illusions; second that the majority get it right; and third, the errors are idiosyncratic and personal, each wrong in its own way, the only uniformity being in the reports that are correct. If the observers by the lake were wrong they were all wrong in the same way. These experiments do not account for that.

(c) If we suppose any one of the trio induced his or her companions by spoken words to create a spurious image at the time, it would have to be the steward asking whether the lady 'saw the man on the lake' because he spoke first. The ordinary meaning of those words is 'a man sitting in a boat', but that is not what the lady said she saw, nor what the masseuse saw either; for her it was not only Bowes that had to vanish but a boat also; besides, people do not ordinarily image anything on a suggestion from someone else. If you spent time among the tourists waiting on Westminster Bridge for Big Ben to strike the hour, asking them afterwards whether they had seen the dead dolphin on the low tide banks below them a few might say 'Is that what it was', and one or two might even say 'Yes I did'; but if you said to them at the time 'Look at the dead dolphin' – when they could make a contemporary check for themselves – they would all see nothing. The human visualising faculty is ordinarily proof against enduring disruption in broad daylight. And no words influenced the steward. The two women may have been subconsciously anxious about the old man's health and imaged the steward's 'the man' as Bowes 'passing over to the other side' for that reason. Perhaps so, but neither woman had ever experienced a visual hallucination before and it still leaves the steward seeing unprompted a non-existent man standing on the water.

(d) Possibly the trio did share some unusual visual experience by the lake but materially misremembered it? No doubt each person would have talked of it in their own circle from time to time (and Miss Godley must have done since her family knew about it), but that would not make their individual mistaken recollections converge on a single account. Rather the reverse, if the original observers are to amalgamate their various versions they must discuss it together. We cannot tell now whether that was likely to have happened – the steward and the lady having a cup of tea together now and again, or he later the masseuse's patient? However that might have been, Miss Goldsmith did not fall in with her employer's views on the absence of a boat. (Slight discrepancies are usual in shared hallucinations; there are other instances.) Then again, two facts underlying all misremembering scenarios are first that there must have been something to misremember, and then that the earlier interpretations should be entirely forgotten. If in this case the mis-recollection was the simultaneous appearance to three people of something they took to be Robert Bowes on the lake, what was this thing that they all forgot? Another fact concerning memories is that although there are instances of spectacular mistaken recollections there are also instances of correctly recalling events in spectacular detail if they were as remarkable as this one was. My interviews with surviving merchant navy deck officers about the sinking of their ships in wartime taught me that memories of unusual events may not be comprehensive but what survives is usually accurate. I do not think uniform mis-recollection of so extraordinary a sight to be likely.

Deranged perception

Accepting these accounts requires that each of the trio should be sane and sober at the time. There is no reason to think otherwise. The lady with a broken leg was combining an excursion in the fresh air with a visit to a sick man on her estate, and writing about him to the doctor immediately on her return. The other two accompanied her in their roles as employees.

In the context of psychoses no one doubts that some people experience hallucinated imagery – in visual or other forms – but it is idiosyncratic. A eminent psychiatrist I was talking to at a trial where he was an expert witness said he could recall no class of shared imagery in any pathological hallucinations except perhaps *folies à deux* in which those afflicted are habitual companions; one is always dominant and imposes his perceptions on the other who recovers without treatment once the pair are separated. Where there is one solitary subject, mentally impaired or not, there is no reason to suppose the visual content anything but internally generated, purely idiosyncratic, and sometimes possibly random. But not with hallucinations of common content shared by healthy subjects.

* * *

Whenever it transpires in any scientific debate that data is unverified or lacking, a decision can be postponed. That latitude does not extend to historical events – there can never be any more observations of Robert Bowes. All the ordinary explanations are improbable and one has to assess their likelihood by calling in aid one's experience, imagining what those lakeside observers might or might not have thought or said or done in each circumstance, and whether their actual behaviour fitted each postulation.

The justification for supposing that the observers were unhurried is that the apparition endured for something like twenty seconds. Picture the steward walking at the head of the donkey holding a leading rein or similar; he catches sight of something on the lake, appreciates it is unusual, looks more intently and concentrates on it, and decides he really is seeing what he supposed. Three to five seconds I'd say, imagine it yourself. Then he acts to draw his employer's attention to it too, by turning to look back at her, catching her eye, and then saying 'do you see the man on the lake?' – three seconds say. She goes through the same checking process for herself, perhaps slightly abbreviated because the steward having seen it might make it marginally less improbable to her – another three or four seconds say. Then "where's the boat – there's no boat – what nonsense there must be a boat and he's standing up in it" about five seconds more. A pause of perhaps two or three seconds by the steward because he'd just been contradicted and then "who do you think he's like – he's exactly like Robert Bowes, the old man" – about five seconds more. It comes to twenty seconds. It may have been seventeen seconds or it may have been twenty-three but roughly twenty is useable evidence in the circumstances – it cannot be proved but that does not wish it away. I see no justification for denying these phenomena on the grounds that an ethereal figure is an impossibility. True, that is what the observers report, but it is easy to see that they are deceiving themselves; what they really see is an internal image and there is little ground for anyone else to doubt that. The question is not 'Was there a figure on the lake?' but how and why did such an image appear simultaneously in three separate minds?

* * *

Here follow brief reports of three other collective experiences to show that such things happen, though not usually so well vouched for. This description of the first is from the nineteenth century:

> (*Mrs P. Naval Officer* $\psi2$) A Mrs P. was lying on top of her bed waiting to attend to her baby in a cot alongside. The light was on, the door locked, and her husband asleep beside her. Pulling herself into a sitting position she was astonished to see the figure of a naval officer leaning forward against the foot rail of the bedstead, his head down and his face in the shadow of his peaked cap. She touched her husband, asking 'Who is this?' Her husband sat up and shouted at the figure, whereat it straightened up and twice called his name "Willy"

in a reproachful tone. Her husband jumped purposefully out of bed but then dithered while the figure moved towards a wall and disappeared; as it passed the light it cast a dark shadow. After searching the house ineffectually the husband returned and Mrs P. asked if it might mean that her brother in the Navy was troubled, to which her husband replied 'Oh no, it was my father'. Neither described the figure while it was present but each of them created an image of a naval officer. (PSPR, 1889)

The second dates from the 1920s:

(*Mr & Mrs Barber ψ3*) A Mr and Mrs Barber returning from a walk both saw a woman going up the path to their front door, and then vanish as if through it. Both exclaimed at more or less the same moment. The figure was largely in grey, with a shawl and bonnet; it was no one they knew. Mr Barber was in front of his wife and saw it turn into their gate, but she only saw it when it was a yard or two along the path.

The details each saw differed; she thought the figure raised its hand to ring the bell, he that it had more colour to its clothing than she saw. Mr Barber reported that it seemed to him, from hearing his wife describe it afterwards, that the figures each saw differed in other respects as well. PSPR, 1922

The third was experienced in Norfolk around 1970:

(*Floating figure in Norfolk ψ4*) Some years ago a friend and I were returning from an outlying village. It was a mild clear night with a half-moon. Coming through an avenue of trees, we approached a large open field with houses on the opposite side of the road. Appearing to float across that field, just above ground level, was the figure of a woman with arms slightly outstretched. She was completely colourless, with the exception of her long fair hair which streamed behind her. We both stopped dead and watched. To our utter bewilderment, she disappeared straight through the wall of the convent school opposite. A man cycling towards us dismounted and enquired 'Did you see that?' We replied we certainly had. He left us saying 'That's the third time, I don't like it, I don't like it!'

...'She' was gliding roughly 1½ feet above ground level. (Green and McCreery).

These three experiences featured common self-created images by two or three people. There is no plausible explanation for any physical reality inducing any of them. The Barbers had no expectation of seeing such a sight in broad daylight, and though they discussed it between themselves later, rather than a common report resolving the minor differences they actually widened. In the first, verisimilitude induced Mrs P. to add a shadow to the unreal figure rather as the masseuse did with a boat for Bowes. The third was seen outside at night but the sight seemed clear enough.

As always, the question is not whence a semi-materialised 'spirit'? It is by what means are self-creations of sane and sober people unified?

Chapter Two

Successive Hallucinations – sightings of "ghosts"

The subject matter of collective hallucinations covers a wide range of topics, but those repeating appearances that we call 'ghosts' are linked to a single theme. Professional literature tends to avoid that term when considering this type of hallucination and does not address the content or specific imagery of the experiences; it is not much concerned with why the figure is that particular one and not some other, but ordinary people conspicuously are. 'Ghosts' is the common term for an idea that is deep-rooted in all cultures and everywhere accepted on some level or another. A ghost is an unexceptionable dramatic or literary device that is understood by everyone as being 'of' some dead person. The difficulty this poses when it comes to investigation is that most witnesses of spontaneous apparitions think they have 'seen a ghost' – what they take to be a rare but natural sight that others have seen – and they leave it at that; they think it external to themselves and are seldom in the least critical or attentive to detail. Timorous people may be distracted by fear.

The best and fullest record of a ghost in England concerns a figure that manifested itself from time to time in a house near Cheltenham between 1882 and 1889. A medical student who was nineteen at the time of the first appearance wrote about it to a friend in London, and then again about later ones. These letters survived, so that when she started keeping a diary of events she had a complete record. She was at pains to keep the affair quiet. Though a seven-year-old brother and four sisters all lived in or frequented the house she never discussed the apparition with any of them until they too had seen it. This was partly to avoid alarming them, partly to retain domestic staff, and partly (by her father when she told him after about two years) to protect the interests of a friend who rented him the house. It is known as Miss Morton's ghost because that was the pseudonym she chose when writing about it, and that is what I call her. (Her real name was Rose Despard).

Here is an outline of first-time appearances to various people: *(Miss Morton's ghost ψ5)*

1. In June of 1882 after Miss Morton had gone to her bedroom she heard a noise at the door an opened it to see at the top of the stairs a tall woman in black holding a white handkerchief to the right side of her face that she at first supposed to be her mother. The figure descended the stairs, Miss Morton followed, but the candle end in her hand went out and she returned to her room.

2. Later that summer a Mrs K, who was the eldest of the sisters and visited from time to time, saw as she was coming down to dinner in broad daylight at half past six in the evening, a figure in black 'with something white about her' crossing the bottom of the stairs and going into the drawing room. She supposed it was a Sister of Mercy seeking charitable donations. No one was there.

3. That autumn at ten p.m. one evening a housemaid saw a figure which by description to Miss Morton was the same one. She was not told then or earlier about any other appearances.

4. Around the 18th of December, 1883 Miss Morton's younger brother and a friend saw from outside the drawing room window a female figure in black seated at a desk with her right hand to her face apparently crying. They came in to see who had called, but there had been no visitors. The time was three fifteen in the afternoon in full daylight.

5. In July 1884 a younger sister saw at eleven fifteen in the

evening what she took to be a widow in black descending the stairs.

6. On 6th August, 1884 a General A. living opposite saw in the garden a lady in black with a white handkerchief to her face who appeared to be weeping. He came across the road to enquire.

7. In the summer of 1886 between eight and nine p.m. a charlady, Mrs T., was waiting 'in twilight' by the garden door for her wages when she saw a tall lady in black with a white handkerchief to her face immediately outside the door. She had heard of the ghost but at the time found the figure so real that the idea of an apparition never crossed her mind; she thought a visitor had somehow missed the front door and went out to help. On following the figure it vanished. Mrs T. said that a new parlour maid who had been deliberately kept unaware of the appearances saw the same figure from an upstairs window. The girl left her job at the end of the month.

It is worth repeating that except for Mrs T. all those people saw the figure without knowing that anyone else had seen it before. Between 1882 and 1889 it appeared about twenty times all told, six to Miss Morton inside and outside the house, once to her brother and his friend, three times to Mrs K, once to the general, and the rest to family and servants. Other experiences were often the sound of characteristic footsteps, which only for Miss Morton were probably linked to the figure because only she looked to see. In one repeat appearance to her the figure reacted to reality by walking around her father. He did not see it on that occasion. Miss Morton comes over as an intelligent and courageous girl. While others cowered behind closed bedroom doors she opened hers to see what was going on. Rather than keep a respectful distance from something unknown, she followed it. The black thread stretched across the corridors may seem to us a little simplistic, but it was an objective measure of the figure's materiality. So also were her attempts to photograph it, which failed because the lighting was too dim for photography in those days. She tried to speak to it, but it would not or could not reply. When she tried to touch it, it avoided her or, if cornered, vanished. It is a feature of the thinking of the times that she never once doubted that it was an external reality. The first appearance to her was at night in poor light. Since she chose to approach the figure it could only collapse for implausibility or seem to recede by descending the stairs. A sceptic could suppose that this was an illusion based on a shadow or similar which Miss Morton was thereafter disposed to image any indistinct visual inputs in the same way, and must further suppose that later she reproduced the same image from different mistaken inputs each time, even in broad daylight out of doors when everything was clear and distinct.

Certainly familiarity may have suspended Miss Morton's disbelief during a half-hour appearance behind the drawing room sofa, which her father and sisters in the room did not see. She was reading and therefore only glancing at the figure from time to time, and it is a pity she does not say what *it* was doing. But her familiarity with the figure does not allow us to state as a fact that on this occasion she was, or might, or must have been, imaging uncertain inputs as the sight she had seen before in other places. Why would that evening's inputs be uncertain? There must have been enough light in that room for people to see easily. Miss Morton may have been more receptive of these influences because she experienced their effect most often – though bear in mind that she used to seek out the figure – but others saw it in circumstances where bad light or suggestion cannot have played a part. Mrs K, the two boys, and General A, all plainly saw it for the first time in daylight, none having heard of it before. Those who saw it at night were more familiar with lower light levels indoors than we are nowadays. The figure was supposed to resemble a former occupant of the house, the unhappily married second wife of its first owner – but no one who saw the apparition had seen that lady, so there is no certainty about that. It is odd that outside fiction, ghosts seem most often to be of people that the percipients do not know; it is as if recognising a real person in such circumstances might constitute an absurdity occasioning the hallucination to break down or never get started. The appearances in this series started and stopped for no discernible reasons; they were most frequent in the middle of the period. In Miss Morton's record of first appearances two were seen by two people in company, the rest appeared to individuals.

* * *

Here are three brief twentieth-century reports of repeating apparitions appearing to more than one person. They are taken from *Apparitions* (Green and McCreery, 1989) and must date around the 1970s and 1980s. An invalid woman who had been moved downstairs during alterations to the house wrote:

(*Black cat* $\psi 6$)...I saw what appeared to be a large black cat jump down from the arm of [a] chair and rush out of the door several times before I told my 19-year old daughter – who said she had seen it too. Several times after that we both saw it together at the same time. We would compare notes as to the route it had taken. Not always from the chairs, sometimes it came from the corner of the room, scuttled across the hearth and into the other corner where the door is – just disappearing there.

We didn't tell anyone – for one reason we were afraid that we should never get any more domestic help if they thought the house was haunted! However, one day the daily help said to me 'I think I must be going daft. I keep thinking I see your cat jump down out of that chair when I come in here' (We have a large black cat). Then later still a new live-in help said 'Have you got two cats? Because I could have sworn one rushed past me in the passage but when I got to the kitchen he was lying asleep' (We have only one cat!). I didn't tell her.

I had never said anything to my husband of our 'ghost pussy' before, but last night I told him of your broadcast and I said I would write to you. He replied "Oh, that cat. I've often seen a black shadow like that in the conservatory, but I've never said anything because I wasn't sure enough of what it was and you would only be nervous anyway!! ...it seems very frightened of people and scuttles off just as cats do when they are startled."

Later she added:

My daughter married and moved about four miles away about seven years ago, and some years' later I said to her "You know I've never seen the ghost cat since you moved". To which she replied "Oh I took it with me. I often see it".

This account is striking for the variability of the apparitional cat's location and behaviour. Four people saw it but for one percipient the image was never fully formed. It appeared sometimes to one individual alone and sometimes to two in company. It moved, and there can be no question of an illusory misconstruing of a real scene because so many appearances demand as many different scenes to be misconstrued. The variety of its behaviour excludes expectation, as does the fact of the daily help knowing nothing of any apparition. It seemed to be attached to one particular person, but appearances did not depend on her immediate presence. There might be something to learn from this case if we could unravel it.

There are a few examples of similar apparent attachments to

people. In the twentieth century another black apparitional cat removed from a Manchester flat to London, and in the nineteenth a white cat from England to Germany, each animal being seen in both its locations by the same two people in company. Those instances certainly allow expectation as an explanation, but when this cat moved to a new location it ceased to be seen at the old one.

The simple directness of this sensible little girl's twentieth-century report is compelling:

(Grey barred cat ψ7) . . while I was a child of about six, seven, and eight I lived in a fairly old cottage with my grandparents and mother. There all three females used to see what was… just a grey barred cat that used to come down the stairs and along the hall and out through the back door. Once over the doorstep it disappeared. It frightened no one, and we used to shoo it out when we saw it. My grandmother disliked it, but I took it very matter-of-factly, even though I realized quite clearly that it 'didn't really exist'. The cat looked quite solid and ordinary, the only odd thing about it was that it wasn't a real one.

…The cat would appear at intervals. We would see it perhaps 3 times a week for 3 or 4 weeks and then not for a couple of months. I would like to emphasise that the above episodes appeared quite normal, I knew quite well that the cat wasn't real but it seemed a perfectly natural thing and about as ordinary as the postman calling.

In Trondheim, a woman visiting the cathedral there had eye contact with an apparition that others had seen from time to time:

(Trondheim nun ψ8) I noticed a Nun sitting in one of the many niches along the wall… The Nun was looking straight at me and I wondered what she was doing there at this time of the

day. I walked along the passage, all the time looking at the Nun and she at me. I thought I would talk to her as I came closer, but when I was 6-7 feet away from her, she faded away and I saw her no more! I must say I was puzzled, but walking into the west end of the cathedral I stopped and talked to one of the women cleaning the church and said to her: 'I thought I saw a Catholic Nun over in the west end, sitting in a niche, but when I came near she disappeared – how could that be?' 'Oh' answered the woman 'we often see her'. And this I have verified by others.

And on 6th March, 2007 I heard this by letter from the daughter of a colleague of my wife:

(*Seated child ψ9*) In 1999 she was responsible for the safety of workmen in an old house near Maidenhead that had asbestos in it. She had heard tales of two ghosts there, one of a little boy who threw a ball downstairs; she had also heard that carpet-layers working late one night had heard and seen this child. She paid no particular attention to the story, never brought it to mind, and was perfectly at ease in the building. Some time later she was carrying a piece of equipment through the house and on rounding a corner in a corridor saw a little boy sitting on the carpet with his back to her; she guessed him to be between two and four years old. She says "as soon as I caught sight of him he disappeared; it was literally a split second thing". She "did not feel panicked or anything, the experience just was".

When she asked the carpet-layer who had seen the child what it looked like he said he did not know because it had its back to him – a detail she had not known until that moment.

In the first report the invalid had seen the figure of the cat twice before the second and then the third person saw it. Three people saw the grey barred cat which the child reported; the bunching of its appearances is curious. In Trondheim the same figure of the nun had been experienced before by the cleaners. In the house in Maidenhead, a wholly rational young woman saw for an instant the same sight that others had seen earlier. In all of them the apparitional figure vanished without being replaced by anything real that could have provided the basis for an illusion. Only in the cat cases was there any movement.

The following response to a recent ghostly sighting demonstrates twenty-first-century cultural attitudes. It was reported by the Society for Psychic Research in 2002: A small company in the north of England had offices and production facilities in a commercial building that was two hundred and fifty years old. One Saturday, a camera recorded a motionless figure in the reception area. When one of the security guards went to investigate there was no one to be seen even though the image persisted and the guard's did not appear. They took the figure to be a ghost. The management consulted the Tourist Board and they suggested that the Society for Psychical Research be called in. The Society's investigators examined the site, interviewed the security guards and then the makers of the television equipment. It turned out that a momentary loss of power could in some circumstances cause the equipment to freeze and continue displaying the last image until the camera's next programmed move (which for that particular camera was no more than the occasional zoom on to the reception desk). That is how the figure had got to be there; he was one of the managers, unrecognised by the guards in his casual weekend clothing. The businessmen had called in psychic investigators to find a ghost, and the investigators called in TV engineers, those who were familiar with the matter taking a more rational view than those who were not. Just as one would expect.

There are academics who are no more critical than those industrialists. Most deliberate and instrumented hunts for physical correlates of ghosts have failed. These ventures start with the report of a psychological fact – the apparition – and continue with the use of all sorts of paraphernalia to detect a physical one. The negative result is held to negate the psychological fact, though really it is negating the quite arbitrary assumption that what was seen and what was sought are physically connected – there is no reason to suppose so, one might just as well try to photograph a dream. Although in trials significant numbers of people introduced to supposedly haunted sites felt an 'unusual experience' at the spot where the figure was said to appear and there was a significant correlation with ambient factors like light levels or magnetic anomalies, no one saw anything. If anything exists in a haunted corridor it is not an ethereal figure but an instigation to create an image of one. Inanimate stones and bricks have never been found imbued with *physical* properties that reproduce visual images of any kind, let alone consistent ones. Nor can the image-generating facility be minutely localized – the invalid's cat was seen in different parts of the house and Miss Morton's ghost in the garden as well. And whatever that facility may be it looks impermanent, despite its sometimes enduring for years. There seem to be fewer ghosts about than there used to be, but better education and the increased scepticism of our times (if any) is not the whole answer. Miss Morton's ghost maintained a consistent appearance throughout, but after a few years it became less distinct and then just stopped being seen.

Summary

Ghosts taking a variety of forms on different occasions have been reported but it is better to confine attention strictly to successive repeats of similar visual experiences because it is they that demand appreciable non-sensory inputs to unify their particular imagery. There is something to be gained from narrowing the field of study in this way, even though near-identical visual ghosts seem to be a minority presentation. That term, as it is commonly understood, comprises a confusing range of different observations. Apparitional figures appearing just once to a single individual are quite common. One may suppose non-sensory causation in all such solitary cases but nothing about them proves it, whence the deliberate exclusion here of all ghosts but those presenting successively and uniformly.

A peculiarity of ghosts of people is that they seem to be presenting an instant or short period of past times in that the figures wear the same clothes at every appearance, or they behave uniformly like Miss Morton's black-clad ghost with that handkerchief. Sometimes their clothing is dated in style, and there is never any suggestion of futurity. The cultural presupposition that ghosts spring from the pages of history allows the possibility of the images being unwittingly fashioned to accord with that expectation. The objection to supposing any of these to be illusions of some indistinct reality is that all the figures vanished without being replaced with any concrete object that might have been misconstrued, a bush supposed a bear is seen for a shrub afterwards. These experiences are not physical phenomena but psychological ones in which different people were, without expectations, seeing the same things in different circumstances and lighting conditions. Unless such experiences are all fictions harmonized by chance, a few localities are by non-physical means capable of influencing a few human minds to create for themselves repetitions of fairly specific images from time to time. There is no more to be said about them for the moment. The enduring cultural supposition that ghosts are the 'spirits' of dead people seriously muddies the waters. Even if we could agree what that means we are in no position to assert any such thing on the basis of what we know of the matter now. The very existence of that idea, together with its attendant supposition that to enjoy independent existence ghosts must have a physically detectable substance, only impairs observation. It is not the peculiarities of ghosts themselves that renders evaluation so difficult but the baseless preconceptions that we ourselves have about them.

Chapter Three

Crisis Cases – distress to distant friends

Crisis cases are the simplest and seemingly most numerous of all categories of information conveyed in non-physical ways awake. They comprise a class of hunches whose content concerns contemporary distress to a distant friend or relative. Where they are visualized they can be distinguished from ghosts by appearing just once and by the subject being known to the percipient. Here is a well-recorded example that was published in 1901 in the Journal of the Society for Psychical Research:

(Mr. Young ψ10) A Mr Young who lived in Llanelli in South Wales said to his sister-in-law one evening: 'There, I have just had an intimation that Robert is dead'. He noted a sense of Robert's presence behind him and the time, 9.40, in his diary. Next day he wrote a letter about it to his sister. Robert (his bed-ridden brother-in-law) and his sister lived in Sturminster Newton in Dorset – as the crow flies about a hundred miles away across the Bristol Channel. She meanwhile had sent a postcard saying just that Robert had died, adding in a later letter replying to his that on returning home with her niece at 9.40 that evening they found that Robert had died at 7.45. This happened in 1895 but in the days before telephones (as these were outside major cities) contemporary written records were much commoner than they are today. Mr Young's diary entry was countersigned by the sister-in-law who had been present, and a niece he met next day wrote to confirm that he had told her of his intimation the day before the postcard arrived. Mr Young's letter was lost but the sister's post-marked card and letter both survived. There is no proof that timing was exactly co-incident because she only wrote at length after hearing

from her brother, but it was close. On the face of it, the information upon which Mr Young's conviction depended seemed to track from Dorset into Glamorgan by unknown means that must be non-physical.

Mr Young's intimation of the death is necessarily the unexplained emanation it seems to be, or one or more of lying, misapprehension, or chance.

The objection to lying as an explanation in this case is that dishonesty is not a presumption in human affairs; it is not a default value. Mr Young had no discernible motive for dishonesty and the three women who vouched for him even less. If dishonesty is to be sustained there must be evidence to establish it or at least to show its utility, but in this instance there is none. A further objection is that Mr Young's diary was what we call a journal, with blank pages where dated entries of any length were made after an event. He says himself that his proved nothing because he started this entry on a new page, and two days' later made a short note in the space before it, so that the accounts were out of sequence. He does not sound fool enough to have been so careless if he was cheating.

There is some uncertainty over the verification of some of the details. The niece in Wales may have remembered correctly Mr Young's remarks but later confused the date of them, and his sister may not immediately have glanced at a clock on learning of Robert's death. A professor Barrett had earlier suggested to Mr Young that he write down any such experiences that he might have, and his sister wrote of the incident:

I am glad you had a presentiment of poor Robert's release... I came home at 9.40, so that was the time you had the impression.

He seems to have had an interest in such things and his sister a belief in them, so their reports may be tinged with enthusiasm, but on the main issue there is no doubt: Mr Young believed that he had received information while his sister-in-law was present, he acted on it, and it was correct.

As for chance, it is quite likely that an apparently terminally ill man would be in the minds of one or other of his relatives, supposing there were several of them, within say fifteen minutes or so of his death if he died during waking hours. Seriously sick people with six or eight concerned acquaintances each thinking of them in the ordinary way once a day would, on average, produce several such coincidences daily in a population where the mortality was about ten thousand people per week, as it was in the United Kingdom at that time. But Robert was no part of any deliberate and directed train of thought on the part of Mr Young, nor was Robert in his mind by accidental association or triggered by commonplace circumstances. It was a spontaneous intrusion carrying with it a sense of conviction that one particular thing had happened. Robert being on his brother-in-law's mind by chance is not unlikely, but unprompted knowledge that he had died must be less so. As also is receiving such knowledge at much the same moment that his sister learned it.

The appeal of chance as an explanation is further diminished by the informational content of these crisis cases. It is never good news – never personal happiness, academic distinction, sporting pre-eminence, financial good fortune. If these apparent transfers of information between separated people were happening by chance all conditions should be represented, but they are not; with one trifling exception, only information associated with danger or emotional stress gets through – a negation of randomness that demands a cause. The exception is a class of hunches that visitors are on their way or an absent resident is returning home; there are not many of them but they do not weaken the argument about chance and content, for the consistency of their subject matter is equally non-random. They are touched on later. Mr Young's experience might have been unique so I looked out a few twentieth-century examples from Louisa Rhine's *Invisible Picture*:

(*Child burned* $\psi11$) A mother's sudden conviction that her young child in another city had been burned; it had.

(*Grandmother dies* $\psi12$) A chilling conviction that a grandmother was dead; she was.

(*Mother's heart attack* $\psi13$) Sudden conviction that a mother in another city was distressed; she had suffered a minor heart attack at the same time.

(*Jack slips at work* $\psi14$) Awareness of something wrong at a husband's workplace, wife went there, a jack had slipped and the truck on it had crushed his hip.

(*Injury in Alaska* $\psi15$) A woman in California felt a sudden need to call her daughter in Alaska; her son-in-law had been seriously injured.

And four more from the nineteenth century given by Gurney:

(*Friend dying* $\psi16$) Conviction that a particular friend was dying at the moment. Written and spoken record made. He was.

(*Cousin Janie is dead* $\psi17$) A five-year-old in Edinburgh says 'Cousin Janie is dead'. Asked which Janie he said 'Cousin Janie at the Cape', meaning South Africa. Janie did die that day from burns suffered the night before.

(*Sister ill in Yorkshire* $\psi18$) Conviction by a woman in Venice that her sister in Yorkshire is ill. She was.

(*Davie is drowned* $\psi19$) A three-year-old in New York State says 'Davie's drowned'. A telegram a few hours later states that Davie and his brother drowned while skating.

A much rarer variant of crisis cases are visualized. A recipient sometimes images the distressed person, though the experience remains uninformative. Here are a few instances. In this first, an ancient one from Gurney, the simultaneity is only confirmed within a day, but the news would have taken months to arrive in an ordinary way.

(*Lord Brougham* $\psi20$) A Lord Brougham was travelling from Sweden to Norway sometime in the nineteenth century. At about 1 a.m. on 19th December he arrived at an inn and took a bath to warm himself before going to bed. While in the bath he saw the figure of a former school-friend sitting on the chair by the bath where he had put his clothes. When Lord Brougham came to his senses he was lying on the floor by the bath and the figure had gone. After he returned to Edinburgh he learned that his friend had died in India on 19th December; they had made a school-boy pact that whoever died first would try to appear to the other, but had since then lost contact with each another.

There was more to this experience than just the sight of the friend, for somehow Lord Brougham got out of the bath and fell without any recollection of doing either. But whatever happened here, the friend was seen, he did die around that time, and Lord Brougham was no longer in touch with him.

Of many wartime examples, this is from the First World War:

(*Airman seen in India* $\psi21$) The subject was in India in 1917 while her brother was a pilot in France. She was busy in the nursery early one afternoon when she felt compelled to turn around, and saw her brother standing there. Supposing he had been sent out to India she turned to put the baby in a safe place, but on turning back delightedly found that he was gone. She heard two weeks later that he was missing after an early morning flight at about the same time that she had seen him. (SPR)

Given the difference in time zones 'early morning' and 'early afternoon' are the same. In this next one from the Second World War, the simultaneity is a little more certain:

(Commissionaire 'the Major' ψ22) In 1942, the commissionaire of a factory was an ex-serviceman nicknamed 'The Major'; a young employee met him on the stairs at about 11 a.m. one day, wished him good morning, but got no reply. He reported the incident at home that evening, including the fact that the man had "the strained expression of a sufferer from stomach ulcers"; later he heard that 'The Major' had fallen on a dwarf fence post the previous evening and had died in hospital of peritonitis at about the time the young man had seen him. (G&McC)

These three examples are typical of the scant amount of information that visualized crisis cases convey. The factual content is slight to non-existent, less even than to Mr Young who at least knew that Robert was dead. Lord Brougham had no idea whether anything ailed his former friend; the sister in India was actually delighted by the appearance of her brother; and the image of the commissionaire only suggested his actual pain. One might suppose that a visual image would add substance to such experiences, but it clearly does not. The non-sensory stimulus was confined to little more than a prompt to image subjectively that person at that moment. The airman's sister knew perfectly well what her brother looked like and could have recalled unaided the image she saw. Proof that she was not being vouchsafed a real sight is that he would in fact have been seated in his aircraft wearing flying clothing including a helmet; likewise 'The Major' would have been lying on his back in a hospital gown.

Lord Brougham's school friend was unlikely to have died sitting upright in a chair at 6.00 or 7.00 a.m., the time in India. Where crisis cases are visualized they seem never to present the figure's actual appearance, nor show it behaving in real ways; all the imagery is the recipient's recollections.

There are a number of crisis cases where the imagery is auditory rather than visual – the distant person speaking a word or two, sometimes a name, less often some non-specific exclamation. The voice is always well-known to the percipient. (Green and McCreery, 1989)

The first peculiarity of Crisis cases is that distance seems to make no difference – a few yards for a woman in Ontario in the next chapter, a hundred miles for Mr Young, thousands for the airman, Lord Brougham, and Janie at the Cape. The next is that the prompt to image or think of the distant person *seems often to be triggered* by that person's distress. Transmission is not a voluntary act, but there often seems to be an implied element of direction to the recipient by the thoughts of the person suffering. There may be a number of reasons why any individual should create an image of an acquaintance, but to do so at a moment of particular significance to the person imaged is what suggests an influence at work. The phenomenon is not that thousands of people call to mind distant relatives; it is that these intimations are of distress in particular, they are contemporaneous, and they are correct. Given the large number of examples, simultaneity and distress and emotional attachment together, make chance an unlikely explanation where the affairs and circumstances of the sufferer are unknown to the recipient.

An insistent feature of crisis cases is their emotional associations. Researchers in this area deliberately designed Xener cards to be emotionally neutral, but this striving for experimental rigour may have thrown out the baby with the bathwater, and eliminated an essential feature of the phenomenon.

It might be possible to test this in a laboratory. Imagine two friends, perhaps children, immersing themselves in a computer game in which one sits isolated in front of a screen that features a human figure with which the player identifies walking through a wood; he disturbs a ferocious bear and has to flee for his life. He comes to a cliff with five caves in it, four are dead ends and one has a narrow exit to the other side. His friend in another room faces a similar screen but he has a broader view, knows which cave has the exit, and wants to save his friend's life by directing him to it. It would be interesting to see what happened.

Crisis cases are the commonest class of non-sensory experiences but they are the least informative manifestation of them so far as mechanisms go. Their imagery, where there is any, is always self-supplied and they are by their nature sketchy, transient, and subjective. Essential features of them – simultaneity and conviction – cannot be subjected to later scrutiny because the timing is seldom accurately established and the quality of the conviction cannot be objectively evaluated.

Crisis cases are the last numerous category of waking non-sensory information to be examined.

Chapter Four

Minor categories of apparitions; reflections, the Virgin Mary, and others

There are sub-groups of healthy people's hallucinations that can be categorized by circumstance or content. With the exception of Marian apparitions none of them are very numerous. Here are one or two examples of each kind.

Informative Hallucinations

Most hallucinations convey little by way of real and new information in pictorial form but there are exceptions. Here are two:

In the nineteenth century an American doctor named Howard was in Arkansas while his wife was in Michigan. He had gone to his room to rest after lunch and locked the door; shortly afterwards he heard someone open it. His wife was standing there, but on turning back after fetching her a chair, she was gone. When he had recovered he wrote to her saying what he had seen. Her reply went:

> *(Arkansas doctor ψ23)* On the day you speak of I dressed myself with the dress and collar you saw in your vision, also the ring, which you have described as perfectly as you could have done if it were in your hand. I felt tired, and went to my room about 11 o'clock, and immediately fell asleep, and slept soundly for three hours. (JSPR, 1890)

Mrs Howard had become anxious for lack of news of her husband; she confirmed that he had never before seen the dress, collar, or ring she was wearing. It does not matter whether or not the doctor had fallen asleep and dreamed his wife's appearance; the issue is the information conveyed.

Also in the nineteenth century, a major in the army saw:

> *(Fishing colonel ψ24)* . . at twenty yards or so a Lieutenant Colonel, who was dressed for fishing and carrying a rod and net, entering the passage leading to the only entrance to the officers' mess. The junior officer wished to speak to the senior one and followed him in, but found he was not there and had not been in at all. Outside again, he told a colleague what had happened, and after talking together for ten minutes or so they saw the Colonel walk into the barracks attired as he had previously appeared, clothing the major had never seen him wearing before. Unbeknown to the younger man the Colonel had been out for two hours fishing some local ponds. (Gurney, vol. II)

Reflections

This is a famous nineteenth-century case whose original report included a plan of the rooms in a London house:

> *(Du Cane ψ25)* The four Miss Du Cane sisters were on their way to bed. When they came to Louisa's room she and Mary crossed it to the mantelpiece to find some matches, the other two stopping about half way across. A little light came through the blinds. Louisa exclaimed as she saw a figure gliding noiselessly across the next room and into hers, a young man with a moustache in dark clothes and a peaked cap. He was of middle height, his eyes were down, and he looked thoughtful, a slight luminescence allowing the details to be seen. The two girls in the middle of the room saw the figure reflected in a mirror on the wall of the adjacent room before it entered Louisa's. It vanished when very close to them. Mary did not see it because she happened to be looking elsewhere when it first appeared, and thereafter was afraid to raise her eyes. It was unlike anyone the girls knew.

The Du Cane's experience was not said to have been written down for about eighteen months, so there is room for recollections to have converged and the details of the figure's appearance may not be very reliable.

Early in the twentieth century, a woman was seen from behind through an open door gazing into a dressing table mirror. When the percipient moved to one side to see the reflected face there was none; the figure had no reflection at all and it disappeared on making that discovery (JSPR, 1892).

Here is an account from the Second World War:

> (Arc-burning ψ26) . . my brother was killed at sea. This happened while I was at work. Part of my trade was to watch a process of arc-burning; to do this I had to look through a dark red glass; and in the reflection of this glass I could see my desk and chair. I had done this same operation thousands of times, and always saw the same reflection. But this time I saw a sailor sat in my chair. I turned and looked, the chair was empty. Three times I looked into the glass, and three times I saw the sailor sat there. Three times I turned and looked, and three times I saw an empty chair. I knew then that my brother was dead. I just knew. I can't, and couldn't then, say why, but I knew. He was an ordinary seaman in the Royal Navy; he and all his shipmates were lost. (Green and McCreery)

At first sight there seem to be different processes to account for. In the arc-burning example, the sailor's image was not sited in the chair where he seemed to be because, on turning to look directly, it was empty; this hallucination was of an inch-high manikin in the glass rather than a full-sized man in the chair. The hallucinated figure at the dressing table purported to be a real person that had no reflection. By supposing all hallucinated images to be internally generated the contradictions resolve themselves. There is no reason why a self-created image should not be of a reflection; it is conceivable that an object and its reflection might be hallucinated together as a single scene – though I have not come across an instance of it – but an object existing *only* as a non-optical internally generated image cannot possibly be reflected in the real world. There is nothing to reflect, no surface on which light can impinge. There is sometimes a slight suggestion that some of these experiences may come about because a spurious reflection seems less implausible than a spurious figure. Dorothy Sayers has her fictional sleuth Peter Wimsey remark, "Cats and mirrors, there's something queer about 'em." (In which connection one cannot help noticing that cats feature in hallucinations more often than any other animal.)

Other Senses and Media

Features of visual hallucinations are shared by hallucinations of other senses too. Self-imaged sounds are fairly well represented, among them auditory hallucinations that in Norway are called *vardøger*; they consist of noises made by an absent resident coming back to the house. Green and McCreery quote a few tactile hallucinations, some incidental to a visual experience (e.g. shaking hands with an apparitional figure) and some free-standing.

This one was precisely informative as to the pain. The man and his wife both wrote detailed accounts and it is often quoted:

> (*Coniston* ψ27) A lady living by Lake Coniston in 1883 was awoken one morning at about 7 o'clock by a sharp pain as of a blow to her upper lip, and she reached for a handkerchief supposing herself to be bleeding. Her husband was out sailing a boat on the lake at that time. When he returned for breakfast she found that, as near as they could judge, she had woken at the moment that he, ducking under the boom as a squall struck, had been hit on the lip in just that place by the tiller. (Gurney)

The second concerns a friend living with a family who used to waken every few months to feel

Cat on bed ψ28) . . a cat on her bed She would switch on the light but no cat was ever there. When her room was being redecorated in September 1972, she moved to share a room with the daughter of the house. The daughter reported herself feeling a cat on her bed three times in the first night – again switching on the light but finding the animal asleep in its box. Some months' later the daughter felt a cat again, and thereafter sensed it every night – usually just its weight but sometimes it curling up against her or kneading her with its paws as cats do. When she tried to put her hand on it there was nothing there. If lying is excluded, both these episodes are what they appear to be – namely sensations induced in an inexplicable way.

In the following case, hallucinated handwriting conveyed real-life information.

(Nanny has bronchitis ψ29) In 1939 a woman staying with her sister had a letter from her daughter including the words 'Nanny is in bed with bronchitis'. She read out the letter to the sister, and passed on the news to another daughter. Reaching home next day she was surprised that the daughter who had written denied knowledge of both the illness and of the words used. The day following the mother visited the Nanny and found her recovering from bronchitis which she had kept from the daughter who wrote for fear of troubling her. On re-reading the letter the phrase was not there, though its place on the page was clearly remembered.

The nanny, the sister, and both daughters all confirmed the circumstances in writing.

Hallucinated images of this kind conveying real contemporary information appear to comprise a distinct class of experience, but there are too few of them to generalize.

Marian Apparitions

In 1981 six children saw on a hillside outside Medjugorje in Croatia a vision of the Virgin Mary. Some years' later, when John Cornwell was researching for his book *Powers of Darkness Powers of Light,* he visited the Bishop of Mostar in whose see Medjugorje lies. The Bishop told him of a compilation from the annual meeting of the French Society for Marian Studies in 1971 entitled *Vrai et Fausses Apparitions dans L'Eglise (True and False Apparitions in The Church)* [of the Virgin] which lists more than two hundred Marian apparitions by Christians everywhere between 1928 and 1971. (Billet *et al.* 1976). Between 1900 and 1993 there were more than three hundred and fifty in the Roman Catholic community alone. Marian apparitions are set in a framework of irrationality but, given that so many are reported, they warrant a brief digression to see exactly what they are.

Nearly twenty per cent of the visions listed in the French publication do not say who the visionary was. A few appeared to huge crowds and a few concerned not the Virgin in person but weeping statues and the like. More than forty per cent appeared to children, singly or with others, the rest to solitary adults. Fifteen per cent were repeated a number of times. Ten per cent of the visionaries were either engaged in full-time religious duties or believed they had just enjoyed a miraculous cure. All the subjects, barring one Protestant in England and a Coptic crowd in Egypt, were apparently Roman Catholics. Some published accounts are wholly uncritical, e.g., someone 'breaking off' a piece of a rock upon which the Virgin had been seen to stand and a spring of water gushing forth. The profusion of haloes, golden rays, crowns, stars, angels and the occasional saint, is embarrassing to read.

There were no cases of the same figure appearing to different people on different occasions, as do ghosts. The juvenile visionaries were all lightly educated country children whose hallucinations were imaged in simple ways appropriate to their intellectual resources. (On the first appearance at Fatima the figure said that all three visionaries would go to heaven but it would cost the nine-year-old boy a lot of rosaries because, being at first unable to see what engaged his sisters' attention, he had sought to shy a stone at it.) Given the widespread credence these apparitions enjoy, it is striking to find absolutely no good evidence at all.

There are smatterings of fraud as well. Members of the local priesthood at Medjugorje orchestrated endless supposedly repeat appearances to one of the children, turning them from a curiosity into an absurdity and creating in the process an income from a world-wide cult that until recently shielded them from being disciplined for profiteering and intransigence.

However, there is something other than religiosity to some of these experiences and, although they have nothing to do with information, having come this far along the track it is worth seeing it through to its end. Collective Marian apparitions seem to share with secular hallucinations non-sensory inputs to harmonize the position, appearance, and duration of each self-created image where there were several visionaries, but thereafter the influence that belief has on perception seems to be at work in them. That they appear exclusively to religious Christians proves that they are essentially subjective.

Bernadette at Lourdes fell into an alarming trance-like state on about half the times she witnessed the apparition. On one occasion she was watched smearing her face with mud and trying to eat it in response to a command to drink and wash at a very inadequate trickle of water. At Beauraing in Belgium in December 1932 – the Church has authenticated this apparition – five children had a number of repeating visions. The last was in the company of ten thousand people. On that occasion a doctor present pricked the children's faces with a penknife and he or another held lighted matches under their clasped praying hands. Neither elicited any response at all, so it is plain that the visionaries were in very deep trances. The same sorts of tests were done on four girls who had repeating visions at Garabandal in Spain in 1961, with the same result. When multiple visionaries develop trance-like states, the implication is that they all end at much the same time, though no reports address that point specifically.

In the year after the appearances at Beauraing there were ten visions in Belgium and two in nearby France against the ordinary annual average of less than four for all Christendom. There was a craze you might say. One of them (at Banneux) was also authenticated by the Church seventeen years' later in 1949, but all the others were adjudged spurious. Up to the publication date in 1993 there has been only one further authentication – in Venezuela in 1976. The church has difficulties with cases like Medjugorje and Garabandal where thousands persist in venerating what it does not authenticate.

The whole concept of Marian apparitions is beset with credulity but initiation of those which are shared could, at least for the first few in a series, be attributed to the mechanisms that have suggested themselves for hallucinating – namely a non-sensory influence of some kind that seems external to each of the visionaries. Otherwise, no information is conveyed in these events and they are of no value on that score. Religious hallucinations differ from secular ones in lasting much longer, and prolonged ones stray after a while into psychological territory quite different from that where they started. They may have an intensely emotional effect on the percipient.

Chapter Five

Mechanisms of Imaging without Ocular Stimuli

What unifies the pathological hallucinations of delusional psychotic states is that they are repetitive for the individuals who experience them; besides the distress they cause it is their frequency that brings them to the attention of the therapists in the first place and no one doubts their existence. Hallucinations that do not repeat are not troubling to the percipient are rare and difficult to study because single experiences of solitary individuals cannot be verified.

Hallucinations experienced by healthy people were first taken seriously in the late nineteenth century. The International Congress of Experimental Psychology held in Paris in 1889 decided to take a census of them across several European countries and 1684 were culled from a sample of 17000 people in Britain. A good collection of apparitions, defined as hallucinations of human figures, appears in G.N.M. Tyrrell's *Apparitions* (Tyrrell, 1943).

A more recent book by Celia Green and Charles McCreery was published in 1975. It too is called *Apparitions* and it includes hallucinations of animals and scenes as well. These later researchers gathered their data by means of two radio appeals, three hundred people responding to the first and fifteen hundred to the second. The reports they publish are brief and mostly uncorroborated; the majority are undated but all were the comparatively recent personal experiences of people living in 1968 or 1974. In all that follows, G&McC denotes reports by Green and McCreery in their book *Apparitions,* and SPR denotes reports from either the Journal or the Proceedings of the Society for Psychical Research – the Psychological Experiences Index gives the full reference.

These representative examples are grouped under these headings:

How these experiences start and stop.
The incorporation of a hallucinated object into real surroundings.
How real objects intrude into hallucinated scenes.
Movements made by subjects while hallucinating.

Beginning and Ending

How does the human psyche shift from basing its visualisations on ocular stimuli alone to basing them upon something quite different? The switching is usually instantaneous. Reports of hallucinations often say that they start 'on waking', or 'on turning', or 'on looking up from' some activity. If we try flicking our eyes quickly from one part of the room to another we create images of the scene at each end of the move, but not of the transit between them. By watching closely we can make ourselves conscious of the intervening blur, but even then it is not imaged. It may be that non-ocular images take hold more easily in the brief natural intervals when our eyes are not providing us with information. Green and McCreery found that about one quarter of all hallucinations begin on waking.

Though starts are usually sudden, gradual endings are quite common. This one ended by fading. In the year 1863, a man woke up suddenly to see a tall lady in a rich black dress looking at him. He was interested rather than frightened, noted many details, and tested his alertness by pinching himself, listening to his watch and taking his pulse. He goes on:

(*Towel horse behind figure* ψ30) After some 40 or 50 seconds I saw a straight white line crossing the figure. I could not make out what it was, till I perceived the apparition was slowly vanishing away in its place, and the white line was the top of my towel on the towel horse behind. Bit by bit the white towel and other dimmer objects in the room came into sight, behind what was becoming a faint mist. In about 20 seconds it had vanished completely. (SPR, 1863)

Here a spurious image disintegrated rather than faded:

(*Cotswold kitchen ψ31*) I lived in a Cotswold house in 1964. About six months before I moved away I saw one night while busy in the kitchen a tall grey-haired woman dressed in a shabby black dress, aquiline nose, pale face, pale grey or blue eyes. A heavy stiff long apron enveloped her. She stood in the hall facing the kitchen entrance. She moved her eyes and head and looked surprised at the kitchen unit and gas stove. When her eyes fell on me, she first disappeared from the head then slowly down until her very thin legs and big men's boots vanished also. I did not realize it was extraordinary until she started to disappear. (Green and McCreery)

Incorporation of Unreal into Real

Here is a case contributed to the international census following the 1889 Paris conference:

(*Russian cadet smoking ψ32*) Around midnight a Russian cadet convalescing in St Petersburg found himself unable to sleep and decided to smoke. He struck a match to find the cigarettes by his bed, and in its light saw his late maternal grandmother sitting on a stool with her elbows on a table by his bed. He dropped the match in terror, but then recovered himself, lit a candle, and found the old lady still there. Summoning up all his *sang-froid*, as he put it, 'I took a cigarette and blew the smoke from it towards the apparition. Imagine my surprise when I *saw the smoke divide on either side of the apparition, as if* encountering an obstacle' (original italics). Then the apparition got up; I distinctly heard the stool being pushed back, backed out of the room through the open door . . .' It spoke a few words to him as it left, and he rushed down the corridor to his mother's room. (Translated by G&McC from the French of the original Census report page 192.

A consistent feature of ordinary dreaming is a complete incapacity to recognise abnormality, nothing is unexpected and subjects are never surprised by what happens. The cadet's astonishment disallows this being a dream.

This is a strange case:

(Uncle at Birmingham concert ψ33) At a concert in Birmingham Town Hall in 1867 a lady in the audience 'saw with perfect distinctness between myself and the orchestra, my uncle, Mr W. lying in bed with an appealing look on his face, like one dying...' The appearance was not transparent or filmy but perfectly solid-looking; and yet *I could see the orchestra not through but behind it* (original italics). I did not try turning my eyes to see whether the figure moved with them but looked at it with a fascinated expression that made my husband ask if I was ill. I asked him not to speak to me for a minute or two; the vision gradually disappeared.

A letter came later to say that the uncle had died at that time.

In this case a percipient's movements while hallucinating were watched by others who were not. The experience is described in the first person at greater length than this:

(Indian officer lifts blind ψ34) An officer in India in the nineteenth century was after lunch taken with a fellow-guest to see alterations that their host was making to his grounds.

After a while a native servant arrived to say that the hostess wished to speak to him. The officer followed the man into the house, where he was left waiting; as no one appeared or answered his call he asked a tailor working on the veranda where the accompanying servant had gone. But the tailor said there was no servant, that the officer had come in alone, and when he riposted that the servant had lifted the veranda blind for him the tailor said no, the officer had done that for himself. Returning puzzled to his friends he told them by way of a joke what the tailor had said, and asked if they had not seen the servant. Both said they had seen no one, the officer had been in the middle of saying something about the alterations when he suddenly broke off and walked back to the house.

The officer's companions confirmed that he was neither drunk nor ill. The light was good and there was no question of his wrongly imaging any reality remotely resembling a servant, nor any reason to suppose that anyone spoke to him. There is no obvious explanation for his behaviour other than his own description – a self-created apparition. Had there been some ordinary need to go into the house, he could have said so.

In these three illustrations a spurious figure appeared for whatever reason in real surroundings without there being anything indistinct that might have been misconstrued as an illusion. Keeping strictly to the reports, the imagery in each seems at first sight to have been based partly on optical stimuli and partly not. How might an apparitional image be combined with an ocular image to create a single image? An obvious possibility is that some unrecognized faculty is superimposing the extraneous figure on a scene visualized in the ordinary way – and that was the conventional and common sense nineteenth-century supposition. Its difficulty at the optical level is that besides a positive hallucination of the figure there should also be a negative hallucination expunging what lies in the visual field behind it. There follows immediately the question how we perceive the boundaries between what is real and what is hallucinated when the figure is stationary, and how we adjust them when it moves.

The alternative idea is that no matter how familiar the environment may be, both it and the figure are entirely and equally hallucinated. Green and McCreery find evidence for this view in the background settings of hallucinated figures differing from the real ones in small details, or being presented not quite from the subject's viewpoint, or being better lit. They have coined the term 'metachoric' to designate the total hallucination of a partly real scene, writing of it in chapter six of *Lucid Dreaming*:

> As far as we can see there are no cases in which the metachoric interpretation is impossible and the conventional interpretation the only one possible. Therefore to retain the conventional interpretation in those cases where the metachoric one is not forced upon us requires us to postulate two distinct kinds of mechanism, the one metachoric and the other 'partial', for what appears to be a relatively homogenous class of experience... By contrast adopting the metachoric interpretation for all cases involves postulating only one basic hallucinatory process behind the whole class of experiences. (Green and McCreery, 1994).

This would mean that those who see ghosts are hallucinating not only the figure but the whole scene. The observers are experiencing a visual image which they take to be a reality in the world around them because that is what visual images ordinarily mean; but if they are not seeing the world then they are receiving nothing via their eyes at all and their interpretation that they are 'seeing' is mistaken also. In a way it makes ghosts even less exceptionable.

Real Intrusions into Hallucinated Scenes

Only two of Green and McCreery's two hundred cases furnish direct evidence that ocular inputs to hallucinations may concern anything more than a static background – that a real moving object not present at first can be imaged and added to that background later .

In the later part of the twentieth century a man was walking from New College to Broad Street past Queen's College in Oxford. He wrote:

> (*Undergraduate on horse ψ35*) ... as (I) came down the road I saw two undergraduates in short gowns, one was sitting on a chestnut-coloured horse with white socks, the other was holding the bridle rein with one hand and had the other hand on the horse's neck. I was surprised to see a horse there and took a good look at it. Just as I came near the group another undergraduate on a cycle with a tennis racket came round the curve very quickly. I shrank back as I thought there would be a nasty accident, but to my surprise the cyclist came through the hindquarters of the horse. Very startled I realized the horse and young men were no longer there. (G&McC)

The two undergraduates' appearance and behaviour were more fully described; a mounted undergraduate was an exceedingly implausible image for the twentieth century.

In an earlier report a cart had added itself to a hallucinated sight:

> (*Cyclist hits Daventry cart ψ36*) In 1896 a man riding a horse along a straight flat road towards Daventry saw a grey-clad cyclist riding slowly ahead of him who had not been there a moment before. They rode along together until they came up with a horse-drawn cart rattling along in the middle of the road. The cyclist ran with great force into the back of the cart

and disappeared. Not a vestige of him or his machine was to be seen. (JSPR, 1900)

The simplest explanation of a sensory optical stimulus modifying a self-created hallucinated image, though not for that reason necessarily the right one, is that the percipient's brain creates a single coherent image out of a mixture of external and internal stimuli which seems to him to be an object before his eyes. Reports of hallucinations being disrupted by real external intrusions are frustratingly vague on the accuracy of the intrusion's presentation before the scene breaks down, but the single image is seamless for as long as the experience lasts. The real and unreal inputs are linked at a level below the observer's awareness. The perceived result is metachoric, and there is no need to 'blank out' parts of reality to make room for unreality. The fact that movements requiring sight can be made while hallucinating proves a persisting visual awareness of the environment that is not used for imaging – a point well made in hypnosis.

Movement

People hallucinating seem to have complete freedom of movement; they can and do respond to apparitions in physical ways. The Russian blew the smoke; the Indian officer's movements and speech were all a consequence of his hallucinating. The family shooed out a regular apparitional cat. There is an account by a delightful Victorian lady who, on seeing an apparitional man standing over her sick mother one night,

> (Towel flapped at figure ψ37) ' . . tried to flap him away with a towel as I had heard that a current of air will make these things go sometimes, but to no avail'.

I did not note this case myself from Bennett, but Green and

McCreery quote it:

> *(Eyes follow figure ψ38)* Miss Leger and I were... in her room with the door open. I was sitting... with a good view of the door opposite, which was shut, when I saw a small inconspicuous old lady, without hat or coat, in old fashioned dress... (she) seemed to fade away into the door of the room opposite me. I looked so tremendously astonished that Miss Leger asked me what was the matter, and I looking at her face knew she knew what I had seen without telling her.
>
> I ought to have said that I watched the old lady moving along the passage to the door. Miss Leger said she could see my eyes moving.

Green and McCreery have accounts where percipients subsequently had no recollection of movements they certainly must have made. This one will stand for them all:

> *(Snake in Australian well ψ39)* A man in Australia who was about to descend into a well to maintain the pump heard a voice saying "There is death in the well". He paused, told himself he was living alone too much, and then went down. Once there he found a snake 'not more than a foot away coiled round the pump'. He next found himself lying on the ground above with no recollection of how he got there.

As did Lord Brougham when he came to on the floor beside his bath after the hallucination of his school friend. Sometimes there is no certainty that the movement was made at all. Did the old lady watching over her mother really flap the towel? Hallucinating images is one thing and hallucinating movements is another, but if a real activity like the Indian officer lifting a blind can be projected on to a hallucinated servant, it is not difficult to imagine that perceptions of one's own movement may be spurious in other ways.

Expectation, Plausibility and Conviction

Experience appears to be the very essence of the process of ordinary everyday imaging. A woman received a letter from a lawyer saying a cheque was enclosed; being unable immediately to pay it in to her bank, she checked two or three times during the day that it was safe in its envelope. Next morning it had gone. In fact the lawyer was mistaken, the cheque had not been written, and in hallucinating its marbled face and inked manuscript the recipient can only have had been repeatedly imaging her expectations (JSPR 1931-32).

The pressure to hallucinate plausibly is evident in Miss Morton's ghost moving round her father, in Mrs P seeing a shadow, the Russian's cigarette smoke parting round his apparitional grandmother, and the masseuse's need to place Robert Bowes in a boat. It might also account for the scene breaking down when its continuation would be absurd – the mounted undergraduate, the bicycle near Daventry, Miss Morton's ghost disappearing when about to be touched and the figure the Du Cane girls saw dissolving when it approached too closely. There is a suggestion that figures reported to pass through walls may be vanishing just this side of them. Implausibility may trigger breakdown of hallucinations more generally.

Conviction can affect imaging. Psychologists find beliefs – e.g., in unidentified flying objects (UFOs) – influencing the process of perception, a tendency that looks rather like an extension to the ordinary pressure to create plausible images, but one governed by a presumed and idiosyncratic plausibility that is faulty. The Marian apparitions are strongly influenced by belief.

* * *

These hallucinations are associated with everyday people and their ordinary concerns.

They do not repeat.

The percipients are observers of the scene and react to them as they will.

They demand no therapy.

Healthy hallucinations seem to be emotionally neutral.

And these are the mechanisms that suggest themselves:

Whenever an unreal figure appears in a familiar setting the whole scene is metachoric – all of it hallucinated, not just the figure alone.

The information from which the visualising faculty of the brain constructs the hallucinated background is received optically in the ordinary way; the information specifying the added image is supplied in some other unidentified way. Nearly always the two combine to produce a single wholly hallucinated scene.

Details of hallucinated images accord with experience.

Eyes may follow a moving hallucinated image just as they do a real one.

An existing hallucinated scene can accommodate a new reality intruding – as with the cyclist arriving when the man was hallucinating the mounted undergraduate.

Hallucinations usually start suddenly.

Hallucinations usually end by collapsing, or fading, or the figure exiting the scene as an actor might – through a door, say.

They may just stop, as though an ordinary untroubled mind can only accept nonsense for a limited time.

Where they collapse in response to a perceived implausibility, the degree of implausibility that can be tolerated by any particular person seems to be subjective and governed by individual expectation, belief, credulity, and so on.

During hallucinations, a subject's eyes provide information for purposes other than imaging, such as walking about or balancing on a bicycle. There is more on that in the next chapter on hypnosis.

Chapter Six

A Brief look at Hypnosis

The value of hypnosis to the study of non-sensory imaging is threefold. First, it can be induced at will both as to timing and to subject matter. Next, it provides proof that in healthy people optical stimuli can be overridden to produce visual images induced by a hypnotist's words. And lastly, in the nineteenth century there were recorded a series of observations of information being conveyed from one individual to another in a wholly unaccountable way.

To judge by the introductory chapter to Assen Alladin's book *Hypnotherapy Explained*, current hypnosis research is largely slanted towards therapy. Where it is not, the emphasis is on the neurophysiology of the phenomenon. A brief history of the whole subject is given in Kenneth Bowyer's *Hypnosis for the Seriously Curious*. Not until the mid-1960s did psychologists (mostly in the United States) begin publishing non-therapeutic research that was much different to, or better than, anything that had gone before. Peer scepticism led them to the most meticulous experiments giving insight into the nature of the hypnotic trance state. One useful novelty was introducing control groups of simulators into the experiments, which at first not even expert hypnotists could distinguish. This forced recognition of subtleties in the condition that would otherwise have been longer overlooked, and simulators were still used long after the original need for them was over (Bowyer, 1977).

Hypnosis illuminates the real, but ordinarily little-evident phenomenon of inducing the visualisation of things that are not there. Bowyer writes that the de-coupling of comprehension from sensory organs, which is an essential feature of hypnosis, is not complete severance. In experiments with analgesic hypnosis subjects were sat in a chair with one forearm immersed in a trough of icy water which a screen prevented them from seeing. This is ordinarily very painful, but although they felt no discomfort or distress they remained aware that something unusual was happening. In a similar vein, of nineteen subjects who could see simultaneously at each end of a baton a real and a hallucinated Christmas tree light, eleven could distinguish them. And a subject hypnotised to be blind to a particular chair in a space she believed to be clear nevertheless avoided it when walking around the room; random controls asked to simulate hypnosis in that experiment thought the pretence required them to bump into what they were not supposed to see. The hypnotised subject was using her eyes for something other than imaging. Anyone awake and alert who is presented with an interesting image he or she does not understand concentrates his faculties on it until he does – such as unconsciously moving his head to change the viewpoint and check for parallax – but deeply hypnotized subjects do not do this. They accept uncritically whatever appears before them – as examples, a subject told to visualize a person already present is not fazed by the duplication, nor by seeing the pattern on a chair cover 'through' a hallucinated figure sitting on it.

It is possible for a hypnotist to make the subject create visual images of what he is told to see. One of the earliest investigators of hypnotism was the Frenchman Hyppolite Bernheim. His report of this case (at considerably greater length) was first published in English in 1990. A significant feature of it is that the girl seemed to have held the instructions in her mind and started to visualize the flower only when particular and defined circumstances arose, i.e. her awakening next morning:

(*Bernheim hypnotises girl* ψ40) I recently hypnotized a remarkably intelligent young girl... whose good faith I can guarantee. I made her see an imaginary rose when she awoke.

(She saw, touched, smelt, and described it, and then, knowing what Bernheim was about, asked if it was real or imaginary, saying it was impossible for her to tell the difference.) I convinced her it was imaginary but in spite of that she was quite certain that she could not by herself make it disappear.

Bernheim found that for hypnotized subjects, prisms double the images of real objects but not of hallucinated ones – a neat demonstration for its time that spurious imaging is wholly internal and free of the laws of optics (quoted in Bowyer 1977).

In a twentieth-century experiment a bright red light and a number in dim pale green light were simultaneously projected on to a white screen. The number was invisible to the naked eye but showed up plainly through a green filter. A hypnotized subject, who was familiar with coloured filters, on holding to her eye what she had been told was a green one saw the red screen as black (as she ordinarily would), but not the number. In fact what she held was not a filter but plain glass. She was imaging what her brain was telling her, not her eyes. Exactly the same point was made to the American Association for the Advancement of Science in 2002 by Professor David Spiegel of Stanford Medical School. He monitored the blood flow to the fusiform gyrus – the part of the brain involved in processing coloured images – while the subjects were watching a computer display that could be monochrome or coloured. The control group could see the colours whenever they were being shown, but irrespective of the display the others could only see them when the hypnotist said they were there.

Hypnosis provides a dramatic insight into the transfer of information. Esdaile was a surgeon in India in the middle of the nineteenth century who operated on patients under hypnosis before chemical anaesthesia had become established. Gurney reports an experiment of Esdaile's in which he hypnotized a blindfolded patient and then had an assistant put various substances in succession into Esdaile's own mouth – salt, sugar, lemon juice and so on. *(Transfers of taste ψ41)*. In this deliberate investigation the patient apparently tasted all of them. A professor J. Smith of the University of Sydney found the same thing in a trial on a young male subject that was apparently one of a series. Gurney gives single-instance accounts of similar trials producing the same results. All date from the nineteenth century; some seem to have been deliberate tests of a phenomenon for its own sake, but others were more in the nature of an after-dinner pastime. Gurney's reports came from these individuals: the Reverend Andrew Gilmour of Greenock, a Captain Battery of Enniskillen, a Doctor Elliotson, a Doctor Lee, the Reverend Townsend and a Mr Kegan Paul. Some of them included light pricking or tickling, and one *sal volatile*. Dr Sheldrake writes in *The Sense of Being Stared At* that Darwin's contemporary Alfred Wallace when a young school master found the same taste effect with hypnotized pupils. There are in all ten separate reports, including one whose source I have mislaid that concerned a London therapist who was accustomed to use the transferred taste of salt as an indication that his patient was truly hypnotized.

These investigators report the astonishing suggestion that the psychological correlates of sensations may be directly trans-ferable from mind to mind without the mediation of the recipient's physiology. If that stands scrutiny our general ability to receive information by non-physical means is proved.

Chapter Seven

Evaluating Non-sensory Information received Awake

The foregoing representative examples of non-sensory information give the flavour of it but not its scale. Among the
numerous collections of these experiences are Green and
McCreery's *Apparitions,* Louisa Rhine's three books *The Reach of
the Mind, Hidden Channels of the Mind,* and *The Invisible Picture;*
also Dr Keith Hearne's *Visions of the Future* and *The Dream
Machine,* Myers Podmore and Gurney's *Phantasms of the Living,*
and the *Journal* and *the Proceedings* of the Society for Psychical
Research, the last three available online. There are other collections also. The phenomena are reported as external entities
lodged in the real world, but that is no more than an interpretation based on our well-founded expectation that every visual
image reflects a real object outside ourselves. The actual observation is of an image internal to the mind of that observer and its
existence is undeniable by anyone else. The substantial amount of
the evidence precludes denying the *fact* of these subjective and
personal images. Whenever such an image is seen by several
people at once its stimulus must be external to all but perhaps
one of them, and its conveyance non-physical.

Multiple images of common visual content in collective
experiences must be either a common response by all observers
to a single external influence or, if one person can impose
spurious imaging on others without their being aware of it, then
perhaps his or her companions might be receiving their external
influences from him. The intimations of distant circumstances,
occasionally detailed more often just general distress, are
likewise arising from external influences. The mode of
conveyance of these common images, whatever their source, is

quite unrecognised as is that of the intimations of general distress being suffered by a person at a distance.

Kant coined the term 'noumenon' for an idea or thought as opposed to a phenomenon which has material and physical attributes. It is a useful concept in connection with those transfers of tastes under hypnosis. To be quite clear about them, the physical activity of the hypnotist's taste buds and brain give rise to the noumenon that is his recognition of that flavour; that noumenon then conveyed to the recipient where it generates the physical phenomenon of activity in his brain, just as happened with those subjects who experienced non-existent colours in Spiegel's experiments.

In all these inexplicable experiences, visualized or not, there was at first no information in the observers' heads and then there was – it had been conveyed there somehow and its arrival was marked by brain activity that is physical for being detectable on an electro-encephalogram – a physical effect without a physical cause. It cannot be over-stated that non-physical information is only a matter of observation by individuals. The observations may not be as good as one would like, but there they are.

The sceptical approach to them is denial of the evidence on the grounds that the phenomenon is impossible and the observers therefore ignorant. It asks of the effect not 'What is it?' but 'Which recognized process must it have been?' – a course that is only rational if one knows everything.

It is perfectly safe to contemplate an idea as fundamental as this unrecognized information if one chooses to, provided one loads it with no preconceptions. It conflicts with nothing we know already and its having been overlooked is easily accounted for by the fact that it cannot be measured.

Chapter Eight

Information at Large – inputs to minds, computers and industrial processes

What exactly is this abstraction 'information'? The word has collected a range of meanings over the centuries and is used for different things in different contexts. Its root is the Latin verb *formare* meaning 'to give form or shape, to discipline, to arrange' or 'to bring into existence'. That spawned *informare* meaning essentially 'to form the mind by imparting something to it'. In modern English the word means also physical inputs to any inanimate system embraced by 'information technology'. Information is ordinarily imparted to minds in one of two distinguishable physical forms: the first is symbols, the second signs. The distinction between the information itself and the symbols conveying it is clearly illustrated by present day computers. The smallest units of information they deal with are binary bits carried by physical elements that are numerous, identical, and small. They are confined to one of only two alternative states written as 0 or 1; which state it is determines the information each carries. Next up from the bit is a byte that comprises eight bits significant for the order of its 0s and 1s. Bytes stand by convention for specific numerals or letters of the alphabet. In informational terms, letters connect the working principles of the machine to the comprehension of its user.

As symbols, letters differ only from those electronic elements by the information in them being expressed by their shape or their form rather than by their state. Their order by convention forms words, as in *leak/kale* for example. Words are the next level of symbol upwards but they carry little information until they are grouped into sentences – once again in a conventional order determined by the language in use, e.g., in English *the cat bit the dog/the dog bit the cat.*

Language is the ultimate stage in this series of symbols expressing information, but its connection to the levels below it is still conventional. Flip the relationship of letters to words and the same set of facts could be expressed in Finnish; another relationship of bytes to letters might present it in Arabic. A trinary rather than a binary code might be more efficient but would work just as well. Any single body of information can be presented in different and wholly conventional ways. In speech, phonemes substitute for written syllables but they are their equivalents. Numerals are likewise significant for their conventional order and the rules which govern their relationships.

What we have is a hierarchy of symbols. Computer bits with a diversity of two form bytes whose diversity is two hundred and fifty six (2^8). Some of those bytes equate to letters whose diversity expressed as words in English (counting plurals and verb forms separately) approaches a million. Those words can be combined in uncounted ways to express anything you like. At every level there are symbols whose meanings have no relation to the essential physical form of their expression. The meanings attaching to those symbols are the information they convey.

As a simple illustration of the distinction between meaning and symbol, suppose a man were to write a long poem on a computer it would exist as a word-processing file within it. There is a patch of disk space whose physical elements have adopted a particular pattern of 0s and 1s which represent the information comprising his work. That pattern can be shifted around inside the machine or downloaded out of it, but what moves is the pattern carrying the information, not the physical elements expressing it. The pattern is clearly a separate distinguishable entity, although it depends for its existence absolutely upon the storage device holding it. If that device with the only copy of the poem were, after the old man's death, accidentally thrown on a bonfire of his household rubbish then that is the end of the poem. It is as dead as he is. The information expressed by the poem was something other than the physical elements that held it, but without solidly physical expression it is no part of the physical world we inhabit either. Kant's noumenal/phenomenal distinction holds for the poem and its expression.

Marks or signs are different; they are not symbols but they also convey information. Imagine a meteoric impact crater five hundred metres across that formed, let us say, fifty million years ago in Central Asia. When nomads first arrived on the scene they might have noted its regular shape, but any myth they may have concocted to explain it had no factual basis and added nothing useful to their minds. With the advent of modern geological ideas and instrumentation a researcher visiting the formation and deploying his gravimeters, corers and magnetometers and so on, might discover its cause and publish his findings in a professional publication. The signs in the case – its shape, dimensions, the rock fractures, traces of ejecta, seismic returns, gravity profiles, and so on, are the physical consequences of meteoric impacts only. They are physical facts, not abstract symbols. Once recognized and rightly read they allow one single definite chain of physical events and processes to be deduced, and the sole message they can convey emerges naturally without any need for conventions. The geologist sees the signs, makes his measurements, devises an explanation for them, and only then reduces it to printed symbols to inform other geologists. The abstractions that constituted the geologist's ideas found physical expression as symbols on paper, but what was their form while still inside his head before he wrote, or in the heads of colleagues after they had read his words?

The concepts in every mind depend upon physical brains, so concepts must have physical correlates, but are those correlates symbols? When one bit in a computer flips from 0 to 1 or back again it is inert immediately afterwards and remains passive whenever it is read. By contrast a neurone always needs a low level of activity to keep it healthy. Though the workings of a brain are exceedingly complex, one might say as a simplified generalisation that when synapses fire they set up a number of electrical signal pathways within and around their neurones, and all these are integrated before presentation to the output axons which, if the resulting stimulus is above a threshold, pass on a pulse to the next neurone. Walter Freeman, who worked on perception at Berkley in California, showed that recognition of particular smells by animals does not relate to small static arrangements of brain cells in fixed locations as on a computer disk. Odours trigger patterns of lobar activity that he detected on a sixty-four electrode array arranged in a four millimetre square. When a new smell is perceived to be important (rather than just experienced) the creation of its peculiar activity pattern modifies all the others, even those entirely unconnected with it. He found that the pattern associated with any particular scent is recognisably similar on each occasion but not identical, rather like some people's signatures (Freeman, 1980). A lot of Freeman's work concentrated on olfaction in rabbits, but investigations in other animals, and of other senses in less detail, showed the same principles at work. The creature's awareness depends not upon a single specific and microscopic brain site responding to a given input from sense organs, but upon patterns of activity affecting wider areas that reflect the meaning of that input.

That activity is the firing of arrays of neurones. The electrical and chemical activity of such arrays are physical processes distinguishable from the concept they represent and to that extent may be regarded as symbols, but cerebral symbols are extraordinary things. Printed words inform by their unique and unchanging appearance; the significance attaching to the elements forming bits in a computer is their present state *vis-a-vis* the other possibility; in brains the symbol with meaning seems not to be a fixed shape or state but a process, a continuing activity. What carries the information is the effect of alternating electrical currents whose amplitude and frequency in an individual neurone can vary, those variations presumably accommodating a wider range of significant differences than anything static can.

Whatever the precise workings of networks of neurones, for humans they make a third category of symbol. The first is those that are immutable like printed words or drawings, next digital computer patterns that can be re-cast, and then arrays of neurones that have the further peculiarity that they seem capable of being modified and even created by themselves. The shape of the letter 'A' and the arrangement of the bits in its byte have been devised by human beings; but I have no idea how any particular array of neurones is supposed to have got its spatial configuration.

The spatial patterns of cerebral activity in Freeman's rabbits were reflecting the meanings of the signs their senses detected, and all such signs – for humans, rabbits or anything else – are confined to sensory stimuli that are *caused by the physical presence* of an external object or circumstance, and their meanings are those learned by experience (or perhaps by inheritance) that relate to the creature's immediate well-being. A scent is laid down by its originator's actual presence, as are its spoor, its droppings, and the leavings from its food. Those marks are not abstract symbols but a physical signs of the animal having been there; in the same way as that crater was a sign of a meteorite having fallen.

Information received as signs may be called 'consequential' information in the sense that its triggering inputs derive from the physical consequences or material residues of that which the information concerns. Information being conveyed to human minds as symbols through sight or sound (or in Braille touch) may be called 'representational' information – the symbols are not caused by the subject matter, but they represent it. Besides signs and symbols both humans and animals appear to receive

information via non-sensory inputs that we have yet to understand. It looks like independent category of information because
it has no recognisable physical basis and it is therefore necessarily without symbol or sign. We humans receive three kinds of
information – 'sensory' via all five senses, 'representational' via
three of them, and non-sensory we know not how. With limited
exceptions animals receive only the 'sensory' and the 'non-
sensory'. That non-sensory influences affect the brains of lesser
creatures than *homo sapiens* suggests that the faculty pre-dates us
in evolutionary terms and that this kind of information is a long-
standing natural feature of this world.

The meaning of the term 'information' is obvious for the
human brain for which the concept was coined, but it has been
extended to physical applications as well. One would unhesitat-
ingly call an input of temperature to a computer programme
regulating some industrial process information, and the actual
temperature value itself at any given moment only the subject
matter of that input. If by mistake the sensor had been left
disconnected from the computer those fluctuations would still
continue, but they could effect no change; they remain just as
real as they were, but they are no longer informative. Had there
been, besides that direct connection, a moment-by-moment
recording of the changing values that conferred temporal
endurance upon them, they might later have affected a thought
process rather than the production process, and have been information with a different effect. Similarly, if through a glitch in any
software, a particular field were not called down in some
computed calculation and the programme comes to a halt, the
field still exists and could be accessed later on.

To be reckoned information any input must both exist in an
appropriate form and endure for long enough to effect a *change*
in the physical world. The contents of a sealed envelope have no
influence on anything and are at the best potential information
only, because the envelope might be lost or destroyed unopened.

That meteoric crater was likewise potentially informative also because it could only inform when observers knew enough geology and astronomy to recognize it for what it was; had it and the area around it been subsumed into a volcano a million years' later then all the signs of that vast physical event would have been erased, and with them the information they had carried.

In a similar way, symbols degrade with time. A symbol, however permanent in itself, may outlive its capacity to inform. An undeciphered script – like Minoan Linear A – is a simple example.

The consequence of the arrival of any information is in one way or another an irreversible re-arrangement of matter or energy effecting a change that is both immediate and physical. The re-arrangements associated with potential information, like the unopened envelope, are not immediate and need not come about at once; and in a similar but not identical way with that crater, materialisation of the information as thoughts is delayed until the receiver's perception alters in an appropriate way that allows him to understand what he sees. All signs and symbols, including those cerebral processes that are the symbols with which we think, are firmly in the familiar physical world.

Information is a noumenon but the various means of its conveyance are all phenomena, excepting only those non-sensory and non-physical inputs to minds and brains considered in these chapters. That said there has appeared in the last fifty years or so inanimate matter apparently being influenced non-physically also. It is very remarkable. The next chapter introduces it.

Chapter Nine

Non-physicality on a very small scale – single particle interference

The word 'quantum' is often used as an adjective pertaining to effects of significance only on a very small scale. They are entirely outside our everyday experience and there is much still to be understood about them. Quantum meaning in loose lay terms "mysterious but respectably believable", is sometimes unjustifiably used in parapsychological contexts to foster credibility. For that reason I am avoiding it altogether; I am not a physicist and do not pretend to an expertise I do not have.

What follows is an outline description of the peculiar behaviour of matter on an extremely small scale. It is a repeatable experimental laboratory effect that looks quite unrelated to the information in psychological contexts that featured in earlier chapters, but there may be parallels. Originally known as 'Young's Experiment', it is remarkable for

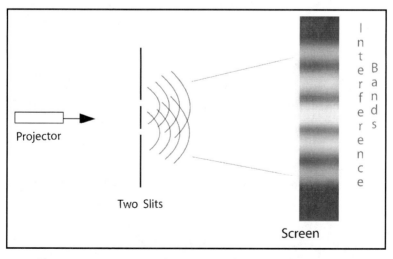

Figure 1: Young's Experiment - Single Particle Interference

two reasons. The first is that Young was an astonishing prodigy who first performed it more than two hundred years ago - in 1801. It was so far ahead of its time that it was forgotten until done by someone else about fifteen years later. The second is that at the time it proved the then-disputed wave nature of light, but more than a century later it was found to demonstrate something else at least as fundamental.

The original arrangement was a narrow light source in a darkened room shining on a screen with two closely spaced vertical slits cut in it, and behind that a second plain white screen. There appears on the second screen vertical bands of light and shadow.

This is an interference pattern, the light waves passing through each slit reinforcing or cancelling each other. If interference of this kind is an unfamiliar idea, picture sea waves bursting through two small closely spaced gaps in a breakwater; they would radiate as semicircles, each centred on one of the gaps; where a crest from one pattern met one from the other the water level would be raised, where a trough met a trough lowered, and unchanged where a crest coincided with a trough. That is what happens with the light in Young's experiment, but it produces bright and dark bands rather than higher and lower water levels. It has been repeated with variations thousands of times and works with light as well as streams of electrons which are easier to visualize. Essential to them all is the equivalent of two small apertures with a suitable detector behind.

Modern versions of Young's experiment use beam splitters instead of slits. These devices send the photons or electrons along either one of two separated paths that converge again near the screen to interfere as before. The value of separating the paths is that it allows room for instrumentation along each one. Single electrons can be fired through the apparatus one at a time and detectors can be placed on each path to see which one the particle transits; the great curiosity this brings to light is that when the detectors are on *the interference vanishes!* Instead of the bands you get a blob of impact marks behind each slit. An inter-ference pattern thus appears not only when there is a choice of two routes, but critically and as well upon the experimenter *not knowing* which one was used each time. If he does know he gets only two blobs or peaks instead of several. The detectors are not themselves disturbing the outcome in some way; if only one path has a detector on it every electron that leaves the transmitter (or arrives at the screen) without triggering that detector must with certainty have used the other path unimpeded as before, but you still get two blobs. Common sense dictates that an object with mass must use one or the other path but it behaves as if it had used both.

Figure 2 shows a Japanese version of the experiment done in 1989 by A. Tonomura and his colleagues, published in the *America Journal of Physics* (Tonamura A et al. 1989). Though each individual succeeding particle seems free

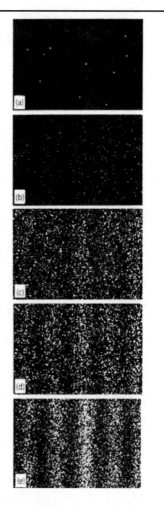

The interference pattern built up by electrons fired singly at two close vertical slits. The frames shows these situations:
    (a) after ten electrons have
         been fired (hits exaggerated)
    (b) after 100, (c) after 300
    (d) after 20,000, and
    (e) after 70,000 hit the screen

Figure 2: Tonomura's Experiment

to go anywhere, collectively they congregate into interference bands. It raises the question, with what is a single particle interfering?

Responses to these findings as they unfolded were varied and original. At first, when the problem was accounting for a single electron producing a pattern at all, some imagined that a real electron has a ghostly doppelganger which goes through the other hole and arrives at the recorder to create the interference pattern pertaining to two. Others supposed that somehow the electron 'knows' whether the path not used has its detector switched on, and responds accordingly. There were other suggestions as well.

Since the fact of observation was changing the outcome some, to start with, inferred that the observer's brain was itself part of the apparatus, but it is not. Suppose in a particular run electrons were fired ten times a second for three minutes, 1800 of them in all, their firing and passage along one or other path being recorded by a computer. At that rate no observer could register any individual electron in real time, but could afterwards examine the track of any one that he or she chose.

One might arrange for the recorder to be connected by a shielded cable to the computer half a mile away so that the signal indicating an electron's passage had no physical connection to the beam-splitter or the slits and none to the screen either; and it could be arranged further that after every ten runs the path records of five were randomly and automatically erased before anyone could look at them. There would still be no interference because at the moment that mattered, certain information existed, no matter what happened to it later. Certainty makes the difference even if no one avails themselves of it. What determines the pattern on the screen is not actual knowledge of the path used but the possibility of knowing it.

Another way of looking at what is going on is to suppose that, once fired into empty space by the projector the electron has a range of positions it can occupy; and that it only adopts one of them when it arrives at the detector (if there is one) or at the back screen if not. Its arrival makes a tiny irreversible mark on the world in one or other of those locations. Before that the *probability* of its occupying any particular spot is, if quantified, wave-like in form – it is more likely to go almost straight ahead but it can sometimes deviate a little, and rarely a lot. That is why in Tonamura's experiment the brightest band is in the middle and the ones each side increasingly faint. From this viewpoint it is not the electron that interferes with anything but the determinants of its position which themselves have no mass. In a similar but not identical way the mass of the open ocean is not pouring through the breakwater gaps as a river would through a breach in a dam, only the waves pass through while the drops of water follow a distorted circular path that has them moving up and down in more or less the same place. On the evidence of the detectors the mass of the electron passes along one or other of the paths through the apparatus, but if both are switched off its destination is decided by laws that are not to be found in the physical world. If the electron's passing leaves a mark on the physical world then the laws of that world govern where it should go, but if its route leaves no mark then other laws hold sway. The existence or absence of information on its path is the deciding factor.

There is with single particle interference an apparent aberration or derangement of space, the difficulty is explaining the *position* of the particle. The laws of Physics which seemed so all-embracing in the nineteenth century have been found to be less so in the twentieth but, as Richard Feynman once remarked, 'the quantum world is not paradoxical, it's just that our expectations were wrong.'

There is more on this later.

Part II

Reports of Apparent Precognition

This entire section relates to intimations received asleep or awake which seem to presage coming events. The term 'apparent' precognitions is not used every time but that is to avoid repetition rather than a jump to any conclusion. They are a varied phenomenon.

There are notes on the performance of sleeping brains followed by a theory of dream structure generally.

The topic is introduced with instances of foreknowledge that appears to relate to events that have yet to happen, and then to the arrival of news of events that have already happened at a distance, either arriving as a waking conviction or in a dream.

There are comments on the possibility of reporting errors, deceit or chance. Special features of precognitions, like their latency, the length of time between precognition and event, and the problems with intervention to avoid the outcome are discussed, as are whether, in the light of dreams of post mortem events, any future disclosed may comprise purely personal reactions to the future circumstances rather than the fact of the occurrence itself. The phenomenon of 'delayed choice' appears to demonstrate an anomaly with time on a very small scale.

Chapter 10

Premonitions; waking precognitions; a striking dream

What sources are there for reports of precognition?

Four hundred and fifty people (some with multiple examples) responded to a Sunday newspaper article by Dr Keith Hearne in 1982. Thirteen hundred contacted J.B. Priestley after a television programme in 1963 and Dr J.B. Barker collected about seventy on the calamity at Aberfan in 1966. (Priestley 1964, Hearne 1984). Dunne refers to about forty personal precognitions and Mrs Hellstrom recorded about sixty; Saltmarsh noted two hundred and eighty logged by the Society for Psychical Research in its first fifty years and one hundred collected by the French professor Charles Richet (Dunne 1927, McKenzie 1974, Saltmarsh 1938). An analysis by Ian Stevenson of seven thousand accounts of unusual experiences collected over almost half a century by Louisa Rhine in the United States found three thousand four hundred to be precognitive dreams. That is more than five thousand in the twentieth century and about three hundred from the nineteenth.

There is a large body of material collected by responsible people. Dr Hearne is an English psychologist who has done a lot of work on dreaming and lucid dreaming, and significant parapsychological work as well. J.B. Priestly was an English novelist and playwright who became interested in Dunne's work and wrote a book based on cases collected after his television appeal. Dr Barker was a consultant psychiatrist in England. Eva Hellstrom was a Swedish lady who had remarkable dreams and kept a diary of them. Charles Richet was, in the nineteenth century, what we would now call a parapsychologist. Ian Stevenson was a Canadian psychiatrist who was professor of psychiatry at the Virginia School of Medicine, and Louisa Rhine assisted her husband Professor J.B. Rhine at Duke University in North Carolina and continued his work after he died; among her own work was a huge collection of spontaneous psychological experiences.

* * *

Premonitions of Avoidable Dangers

There is a class of useful experiences that comprise intimations of existing situations in time to avoid what would otherwise be their undesirable consequences. The intimation is not of the event to occur but the timely receipt of information that will arrive in ordinary ways when it would be too late.

Jim Corbett was a famous hunter with an unrivalled knowledge of the jungles of Northern India, particularly the behaviour and cries of all the creatures living in it. This skill allowed him to know the identity and activity of every creature around him that had alarmed, or been alarmed by, any other. Over twenty years or so he put this knowledge to ridding areas of man-eating tigers by stalking them alone on foot. One exploit, which is written up in his *Man-eaters of Kumaon* as *The Chowgarh Tiger,* concerned a tigress that was credited officially with sixty human lives and with twice that number locally. On learning of another kill he left for the locality and started searching. He writes:

> (Tigress $\psi42$) One evening… [*he was following a cattle track to his bungalow after searching for a fortnight or so*] …when approaching a pile of rocks, I suddenly felt there was danger ahead… A hundred feet would see me clear of the danger zone, and that distance I covered foot by foot, walking sideways with my face to the rocks and the rifle to my shoulder; a strange mode of progression had there been any

to see it.

Moments after rounding the next bend in the track, a Karker deer gave its alarm cry specific to a tiger and Corbett strode quickly back. In the damp clay of the track by the rocks he saw the prints of the tigress overlying his own. He writes:

> I have made mention elsewhere of the sense that warns us of impending danger, and will not labour the subject further beyond stating that this sense is a very real one and that I do not know, and therefore cannot explain, what brings it into operation. On this occasion I had neither heard nor seen the tigress, nor received any indication from bird or beast of her presence, and yet I knew without shadow of doubt that she was lying up for me among the rocks... I knew they held danger for me, and this danger was confirmed a few minutes later by the karker's warning to the jungle folk, and by my finding the man-eater's pugmarks superimposed on my footprints.

Corbett next met the tigress four or five days' later in the most hair-raising circumstances – glancing behind a slab of rock he had just passed he saw the animal crouched to spring only eight feet away; he froze, and it took the utmost self-control to bring his rifle slowly to bear, one-handed, without provoking an attack. So far as that episode concerns this book, the striking thing is that Corbett had a strong and clear but wholly unvisualized premonition of danger on the first occasion, and on the second – when on the face of it he needed one just as much – nothing at all, though on the second occasion he did get a sensory warning of the danger in time for his own expertise to get him out of it.

Hearne presented two similar warnings of present danger, the first partly visualized. A woman being driven by her husband wrote:

(Tree on Road ψ43) I told him he had to slow down. I could give no reason but in my mind's eye I could see a tree across the road farther ahead. A few yards ahead a fellow was swinging a light in the road and stopped us to re-direct us. There was a tree across the road. (JSPR, 1983)

And a motorist driving to work in East Anglia recorded that as he was approaching a sharp, blind, left-hand downhill bend in a built-up area:

(Dog and pedestrian: ψ44) Above the hedges and between the houses I suddenly became aware of the top of an enormous truck coming up the hill and nearing the bend from the opposite direction... I slowed down and got ready to stop... However, there were no other vehicles at all, but what was there was an old lady crossing the road with her dog. Had I gone round at my normal speed I would most certainly have hit her or crashed the car to avoid her.

The imagery may or may not have accurately reflected contemporary circumstances in the first case. In the second it was wholly wrong but it led to the right action. If, in the second driving instance, the danger had been a real lorry he could not see then the driver might subconsciously have heard it and hallucinated the old woman to instigate his caution; but that is not what happened, it was the other way round. There was no ordinary way he could have known of the danger that the pedestrian presented.

Louisa Rhine also has a few cases where the distant situation was an unspecified danger unknown to, or unappreciated by, the person in it.

(Ontario Snake ψ45) A woman on a farm in Ontario was chopping firewood when she heard, or perhaps felt, the

command: 'Never mind that. Go back to the house!' There she found a large snake staring at her baby who was innocently staring back.

Here the danger was a gun:

(Toddler with revolver ψ46)... a couple out for an evening's bridge with an aunt put their three-year-old son to sleep in an adjacent bedroom; the wife suddenly – 'quite rudely. I didn't even know I was doing it' – leapt up from the card table to find the child playing with a loaded revolver that had been forgotten under the pillow.

Waking Precognition

Most intimations of unrealized events are presented in dreams but a few of them arrive awake. Here is an example of fully visualized foreknowledge experienced by Mrs Hellestrom. Andrew MacKenzie included it in his *Riddle of the Future* (MacKenzie, 1974). The original account is in the Cambridge University library.

(Coptic Rose ψ47) In 1949 she was travelling on a coach from Germany to Italy to join her husband for a short trip to Cairo. Passing Heidelberg in the dark there suddenly came into her mind a clear coloured visual image that she took to be some kind of painting measuring about twenty-five by thirty centimetres. The dominant impression was a dark rich pink background and a central figure like four hearts meeting in the middle. Only half noted was some black ornamentation like spirals running between the hearts from the corners to the centre. She had no idea what it was but somehow expected to be seeing it quite soon, and made a sketch of it in a notebook.

This is a copy of her sketch made in the bus and a photograph of an object she saw later:

The decypherable text reads: *Rosa fält* - pink fields; *Lite* - some, a small amount
Cerice – cerise; *Svarta ornment* - black ornament

Figure 3

The experience lasted for about five seconds. She showed her sketch to the hostess on the bus, to her son in Florence, her husband in Rome, and to two Swedish ladies she met in Cairo. On her last day in Egypt these ladies suggested a visit to the Coptic museum. Entering the second room one of them said to Mrs Hellestom 'there is your image'.

It was carved on a flat stone slab forty-three by forty-eight centimetres, Mrs Hellestrom not noticing it because for her the striking feature was the exceptional colour while for her friend, who had seen only the pencil sketch, it was the outline. The curator told them that the design, known as the Coptic Rose, had been fairly widespread in ancient times. It was usually coloured in the same hue that Mrs Hellestrom had seen; a few tiny flakes of paint still adhered to this slab. Mrs Hellestrom had never heard of the Copts before her visit to Cairo, and no photograph of this specimen had ever been published.

Her description of the carving and the drawing that she made in the moving bus together bear a remarkable resemblance to the reality. The two principal omissions are the circle in the centre and the spiked petals, but the essentials are there. Another curator of the museum later shown the sketch without knowing its history immediately identified it as the design on the slab. Receiving an intimation of the original colour, which for Mrs Hellestrom was only ever associated with the design by a verbal description of its appearance centuries before, is remarkable.

Here is another case included in the same book. In 1935, Wing Commander Goddard who was piloting an open plane without blind flying instruments got into difficulties in low cloud and driving rain. He regained control about fifty feet above the ground and made for an abandoned First World War airfield called Drem to check his position. As he flew over the boundary:

> (*Goddard over airfield ψ48*) . . the whole scene seemed suddenly to be bathed in full sunlight; the bitumen on the newly repaired hangar roofs were glistening with recent rain, several yellow training aircraft of old and new designs were parked on new tarmac and one was being pushed by men in blue overalls. At that time RAF aircraft were aluminium doped and RAF overalls were brown. After the few seconds it took him to fly across the airfield the rain closed in again. When Drem was re-activated four years' later a new trainer called the Magister had come into service, and all training aircraft were painted yellow; RAF technicians had been issued with blue overalls.

Hearne gives this instance of a semi-visualized hunch seen 'in the mind's eye'. It seems not to have been mentioned to anyone before the event, but it led to appropriate action in the present:

> (*Bomber shot down (ψ49)*) In the RAF I flew on bomber raids over Germany. In May 1942 I had a strange premonition that I would not return from the next raid. In my mind (not a dream) I "saw" myself floating down over enemy territory. There was an intense fire and the blazing bomber hurtled

down and smashed into a village, completely destroying a pub. The premonition was so intense that I sorted out my personal papers, etc. Our plane was hit over Holland. We bailed out when it burst into flames and hurtled away in the darkness. I floated serenely down to the ground.

We never knew what happened to the bomber until recently. After obtaining the German statements of the gun crew who shot us down, I have been in touch with a chap in Holland who lived fifty yards from where the burning plane crashed. It completely demolished the pub in the village of Elsloo. He has sent maps and photographs of the rebuilt pub and even parts of the aircraft he still has as souvenirs.

The downing of his aircraft came about quite soon but the detail concerning the pub became known to him at least thirty years' later.

Hearne includes these two that related to one wholly unforeseeable future event:

(Kennedy assassination ψ50) On the Tuesday before (President Kennedy's) assassination I had just put my daughter to bed and had a cup of tea and a quiet sit-down before starting our evening meal. I wasn't thinking of anything in particular when I actually 'saw' in my mind's eye the whole episode of the shooting. I even saw it in colour. It was as if I was a spectator in the front row.

His car had just passed me. There was much cheering and President Kennedy was smiling and waving. Suddenly he slumped forwards. People were running towards his car but security men were keeping them back. I came to, and didn't really know why I had thought it. Three days' later when the pictures came on TV it was exactly, to virtually the last details, my day-dream.

And the next:

> *(also ψ 50)* The day President Kennedy was murdered my husband and I were travelling by car to an RAF base in Norfolk. At about 11.45 we were passing through a village when I noticed a newspaper placard saying "President Kennedy Assassinated". *(They were unable to stop.)* We went on looking for another newsagent. At Buckingham I asked people at a café and at two newsagents about the news but no one knew about it and I began to feel foolish. When we arrived at the base I asked my daughter but she too had not heard anything. In fact, the assassination did not happen until several hours after I had seen the placard.

The first episode was an intimation of a fact of the assassination rather than the scene because the car sped away and no one had time to run towards it; if the time was 6.30 p.m. in the United Kingdom the impression may have coincided with the shooting, which happened at 1230 Central Standard Time. In the second, which preceded the event by about six hours, the woman does not say whether she later saw such a placard in reality; she seemed to have visualized in a personal way the fact to be known later. By some fluke her experience came to the attention of the United States security services, who for a time took it to indicate a non-American dimension to that crime.

Lastly comes this from Dr Hearne's *Visions of the Future* . A young woman wrote:

> *(IRA bomb ψ51)* I had just got off the bus in Whitehall en route for my office. As I turned the corner in Whitehall Place it was clear that there had been some kind of explosion; people were lying on the pavement, with others bending over them. There was a lot of broken glass and blood.
>
> The blast appeared to come from the left. Everyone was

clustered to the right as though thrown or blown that way and the windows on each side of the opening into Scotland Place on the left were the only ones that still had glass in them. I couldn't understand why nobody in Whitehall had reacted to this, or why I hadn't heard the explosion. I then realized that there was no sound at all; no screams and no traffic noise. The scene was quite silent, as though a screen had slammed down behind me. Then normal sounds resumed and the 'vision' disappeared slowly.

I staggered into the office for strong coffee, feeling as shaken as if it had been real. The following day, when a rail strike prevented me from going to work, an IRA bomb went off in Great Scotland Yard just round the corner from where I had the premonition... one person died and 238 were injured.

The explosion happened on 8th March, 1973 in Great Scotland Yard. The young woman had visualized something that had not yet happened.

A Precognitive Dream

In contrast, and by way of an introduction to the topic, here is a stark precognition that I followed up myself.

Travelling from Reykjavik in 2010 I saw on an Icelandair in-flight entertainment, a documentary about the volcanic eruption on the island of Heimaey about ten miles off the southern coast of Iceland early in 1973. It covered the evacuation of four thousand people to the mainland in a few hours, the growth of the fissure to two kilometres in length, the raising of the new peak called Eldfell, and the successful five-month battle to save most of the town from being buried and the harbour itself from becoming land-locked by spraying tens of thousands of tons of seawater on to the lava to slow its advance by cooling it.

It made a marvellously dramatic story, but one detail particularly caught my eye – an eleven-year-old schoolgirl called Klara

Tryggvadottir had several times dreamed of the eruption before it happened. No one had paid any attention to her, her mother telling her that there had not been an eruption on the island for four thousand years and there was not going to be one now. The airline told me who had made that film and its director put me in touch with Klara. I returned to visit her myself.

(*Klara ψ52*) She told me that in November of 1972 she saw in a dream a crack in the ground opening up in the hill called Helgafell above her house and lava coming out over the grass. She took it to be a volcanic eruption that would threaten her home town of Vestmannaeyjar. She said she dreamed similar scenes quite often afterwards – always the opening in the ground in the same place, but occasionally other details as well like shadows on her house cast by the glow, or cracking roads, or galloping horses; one dream included seeing the rocks of a crag on the other side of the harbour called Heimaklettur illuminated by flames. Her anxiety increased with the repetitions; she spoke freely of them to her family, to her school friends and teachers, and to the neighbours. All but one neighbour dismissed her tales as childish imagination. On Sunday 21st January, 1973 Klara was in a car with her father, grandfather and sister when they stopped to look at a view of the sea. They were at the site of her dreamed eruption, and at that moment she told me she 'saw', though awake, the ground opening up and lava coming out, together with cracks appearing in the road. She told her father 'not to stop, to keep going, the eruption has started', but both men were short with her.

At 2 a.m. on the following Tuesday a volcano erupted above the town. A spell of bad weather had kept the fishing fleet in harbour so that all the women and children and most of the men were able to leave the island that night. So many people knew of her premonitions that the press heard of it and a journalist came to interview her after the family's evacuation to Reykjavik. I give his account as evidence not necessarily of fact, but of contemporaneity.

He wrote in *Timinn* on 26th January:

These were the words of a little girl, Klara Tryggvadottir from Vestmannaeyjar, the town of the Vestmann Islands, when we approached her yesterday where she was staying together with her mother and a couple of brothers and sisters at the home of friends in no. 2 Hulduland here in Reykjavik. We had news that she had an extraordinary experience to relate, and we were not disappointed on that score.

This is what Klara said:

"I had been dreaming about this for a long time. It started long before Christmas. And I have always dreamt this occasionally. I also dreamt that Mount Heimaklettur erupted, but then I remembered that it isn't a volcano, there's other stuff in it. I had an absolute hunch that it would start erupting. I told Mummy and Daddy about it. I told everyone about it, but nobody believed it."

"What were those dreams of yours like, Klara? Were they clear and similar to the reality that later occurred?"

"No, they were not very clear. But I always dreamt something rather similar, that the eruption started in Helgafell, where it is now, and there was lava flowing just everywhere. I was always so scared in the dreams, and there seemed to me to be so much fire and lava that it travelled across the whole town and everyone had to escape – to Reykjavik, or just get away somewhere."

"When did you last dream about the eruption?"

"I don't remember exactly, it was a short while ago. But

when Daddy took me out for a ride around Helgafell last Sunday with Grandpa he drove along the road above the hay meadow that belongs to Thorbjorn in Kirkjubae (this road was right next to the eruption site and is now under the lava), suddenly I became scared stiff, just all of a sudden, and I told my Daddy he should hurry home as quickly as he could, since Helgafell was erupting."

"And what did your daddy say? Did he believe you rather more than all the others?"

"No, no, he just laughed and said that I could run home if I wanted to. And Grandpa didn't believe it either. He said I shouldn't be talking such nonsense."

The rest of her remarks were about events afterwards, but her mother added:

It never occurred to those of us that knew about Klara's dreams that they still had to come true. She was always telling us about them, and we did nothing other than try to dissuade her from all that silliness.

There is no doubt that Klara had intimations of the coming Heimay eruption and that no one else did, not the seismologists, geophysicists, or volcanologists of Iceland or any other resident of the island. Had there been any ordinary indication other people might have been aware of at the time it would have come to light in the aftermath of so extraordinary an event, and were there the slightest hint of any possible natural explanation whatsoever – changes to the earth's magnetic or gravitational fields caused by the magma churning a kilometre beneath their feet perhaps? – it would long since have been proffered. Something very odd had happened.

Klara Tryggvadottir has spent her whole life in a small island community. When I met her in Reykjavik she was about fifty years old and not in very good health. She had acquired in her family the reputation of being a 'seer' so to speak. The first example her daughter in Vestamanayjaer gave me was her mother 'knowing' when female relatives were pregnant before they did themselves, and of knowing the sex of the child yet to be born; but she failed on both counts with that unmarried daughter's child. She said her mother had been ill some years previously, and seemed to have 'lost her powers'. For all that, her relatives agreed that she had correctly predicted the date of the eruption of Eyjafjallajökull, the volcano that disrupted air traffic over northern Europe for a few days in 2009. Klara told me that Katla was due to erupt soon – but everyone in Iceland expects that; on previous occasions it has followed Eyjafjalljökull by a year or two, and seismic activity – of which the press in Iceland carries daily reports, just as in other places it does for the weather – was on the increase in that area. A little later she said to me, in a conversational way, that it might erupt tomorrow, and her great niece confirmed that I had understood her aright. Had it done so, I think her family would have taken it as another instance of her gift in action. In fact, six years' later, Katla has still not erupted, and by now her pronouncement is probably forgotten. I could not help wondering whether there might not have been similar retroactive adjustments with some of the pregnancies. Even if not, their evidential value is limited because conception is an expectable event for young married women, and the odds against correctly predicting the sex half-a-dozen times running is sixty-four to one, so that, if the births were spread over some five to ten years it might take a family to produce six grandchildren or great nephews and great nieces, Vestmannaeyjar should be able to furnish several instances by chance alone.

Klara and her family are not naturally critical; they have kept no written record of her apparently precognitive experiences, for which reason I did not enquire into them. Some of the differences between what *Timmin* published and what she said to me may perhaps have been omissions arising from a daunting experience for a child, or they could have accrued little by little in the re-telling of her memories over the years. The bare information that Klara Tryggvadottir received was: 'Vestmannaeyjar threatened by an eruption on Helgafell', and there can be no doubt that she dreamed both the fact of the eruption and its detailed location before it happened.

None of the pictures illustrating that information were real sights to be experienced by her afterwards, except perhaps the flames, nor were they derived from films or photographs that she might have seen later because the scenes she reports were not recorded in her town. Her imaging was entirely subjective to illustrate her understanding of an idea, as is the case with most precognitive dream. But none of that detracts from the stark fact that she knew of a quite exceptional event before it happened.

Chapter 11

Dream Structure

Dreams are the commonest presentation of apparent precognition. The content of dreams is usually preposterous. Vivid and amusing they may be, or perhaps terrifying, but always unreal. Our dreaming mind's outputs are not random vagaries, only parts of the brain are disabled in sleep and it is the inadequacy of those which continue to function that makes healthy dreaming so peculiar. During sleep human brains perform in consistent ways that have been studied and is tolerably well understood. Knowing how dreams ordinarily work is a necessary prelude to evaluating any information that they may sometimes seem to be carrying.

Freud's opinions dominated thinking on dreaming for half a century or more after the publication in 1900 of his *Interpretation of Dreams*. The physiology of dreaming has received some distinguished attention – J. Allan Hobson's *The Dreaming Brain* (Hobson, 1988) is an absorbing history of the subject – but thereafter the study of dreaming as a process distinct from its content made only modest progress for nearly fifty years. After the Second World War a number of centres established sleep laboratories where human volunteers provided insight into various aspects of dreaming. The advances since then are well covered in a good and fully referenced book by Jacob Empson called *Sleep and Dreaming* (Empson 1993). What follows is confined to those few aspects of a large and complex subject which bear upon dreams that apparently carry information. Empson's book is their source.

External Stimuli and Dreaming Speed
Attempts to impose dream topics by showing films just before

subjects go to sleep has not worked well in adults or in children. Stressful films produced more anxiety dreams than neutral films, but not on the topic of the film. A series of deliberate attempts to introduce external stimuli into dreams that had already started, as indicated by rapid eye movements (REM), succeeded in more than half of the instances. In another, where names of the dreamers' friends were used as the stimulus, again about half succeeded, but three quarters of them were by assonance alone, by their sound. For example, Gillian was received as Chilean, Jenny as 'jemmy', Mike as 'like'. Assonance is widespread in dreams of any kind.

Incorporating external stimuli into dreams is quite a complex business. A Frenchman named Maury who was interested in this question had an often-quoted dream in 1861:

> (*Maury guillotined* ψ53) ... he was witnessing horrific scenes during the Reign of Terror before being himself brought before a revolutionary tribunal which condemned him; he dreamed in some detail the execution preliminaries on the scaffold, before feeling the guillotine blade fall on his neck; he awoke to find that the headboard of his bed had at that moment fallen on just that very vertebra. (Quoted from *The Oxford Book of Dreams*)

The supposition that the whole dream story was compressed into the few moments between the fall and his awakening found favour for many years, but it conflicts with studies of dreaming speed and is probably wrong. A dream narrative takes more or less as long to unfold as it takes to recite it – a mass of dream material is not conveyed in an instant. In one investigation of dreaming speed, subjects were woken five or fifteen minutes after the start of a dream in REM sleep and asked which the interval was; about ninety per cent of them got it right. In another, cold water was lightly sprayed on to dreamers who were then woken a while after. Those who had incorporated the stimulus into their dreams showed a very good correlation between the amount of dream incident and the interval elapsed since the spray.

An interesting secondary point arises from those investigations of dreaming speed (Dement and Wolpert, 1958). A subject sleeping face down had cold water sprayed on his back ten minutes into a REM episode; thirty seconds later he was awakened and reported thus:

> (*Water on actress* ψ54) The first part of the dream involved a rather complex description of acting in a play. Then I was walking behind the leading lady when she suddenly collapsed and water was dripping on her face. I ran over to her and felt water dripping on my back and head. The roof was leaking. I was puzzled why she fell down and decided some plaster must have fallen on her. I looked up and there was a hole in the roof. I dragged her over to the side of the stage and began pulling the curtains. Just then I woke up.

The water was clearly incorporated into the dream but it did not feature exactly as in reality. He dreamed it on his head and back when it was actually on his back alone and he had already registered it falling on the actress' face before it fell on himself. It is as if awareness of the water on his back was delayed for a moment while his dreaming mind created a link between what had gone before – the acting – and what had come now – the water.

In a second case, a subject lying on his back had water sprayed on to his feet and legs when he was five minutes into a bout of REM sleep; when he was awakened one minute later he said:

> (*Children spill water* ψ55) The dream started in a room talking to some friends. Then two children came in and came over to

105

me asking for water. I had a glass of iced water and I tipped the glass to give it to them. I was sitting and I spilled the water on myself. (He went on that the children tried to grab the ice and he got cross at their greediness then rose to change his wet trousers; after that the scene changed to a school with many children and soon after he was woken.)

Once again there is the introduction of water into the dream story – the children asking for it and the glass being tipped, before awareness of wetness was incorporated. Dement and Wolpert were using the water only as a marker to investigate the speed of dreaming, and did not publish the other thirteen cases where dreamers incorporated the applied external stimuli. It is a pity we do not know how they did it.

The Marquis de Saint-Denys (Saint-Denys, 1964) noticed this same delaying of sensations. He kept notes of nearly two thousand of his own dreams, the largest pool of personal observations that I know of. In 1867 he wrote a book on them. He illustrated the delay with this imagined dream (my translation):

When bitten by a mosquito while sleeping, his dreaming mind interprets it as a minor sword wound, but it cannot incorporate that into a dream without preparing a context for it. He supposes a quarrel, an exchange of insults, a duel proposed and the arrangements made; finally swords are crossed, and it is only after all these preliminaries that 'I seem to feel a slim sword run through my arm'.

His conception of the process is that the context is created after the need for it but is presented as if dreamed first. Delay while that context is prepared is central to it.

A personal experience of my own seems to show a brief external stimulus that persisted for less time than it took to modify the dream narrative. I was staying in a Corsican hotel outside which a clock in a former church tolled the hours. There were no introductory chimes to announce each strike so it was repeated two minutes later. I dreamed:

> ...of a group of urban terrorists with whom I was personally friendly but politically out of sympathy. The comedian Billy Connolly was one of them and through my chained front door he was trying to persuade me to let them hide in my house. Then the setting became a room full of people being quizzed by a security official, he was asking someone else what I had done on the occasion of a particular hold-up. The speaker said that I had asked the gunman a couple of questions (which he had answered) and had sat down and raised my hands. When the official turned to talking about other people I sidled to the back of the room – and dropped dead of a brain haemorrhage.

At that I awoke, amused by the abruptness. I thought about it for a bit, and turned over to go back to sleep. Just then the clock struck three – but once only, it was the second strike. I suspect that although the first set of strokes went unregistered the last one must have woken me, for otherwise the silence could have allowed me to go on sleeping. My dreaming mind seemed to have coped with the imminent shut down of the narrative by writing me out of the script altogether – once again, as with Maury on the guillotine, in the closing moments. If that is right it had a finite time in which to do so – next day at breakfast on the terrace I timed the clock's striking rate at three in three and a half seconds.

To condense the salient features of ordinary dreaming as they appear at the first look: Everyone dreams, probably everyone dreams about the same amount. Dreaming is continuous in REM sleep and sporadic otherwise. Everyone forgets his or her dreams very readily. Attempts to induce dream topics before sleep do not work, attempts to influence dream content by external stimuli show spoken words being assimilated mostly by assonance rather than meaning, and the registering of physical sensations seeming to be momentarily delayed while the dream narrative adjusts to them.

The Elements of a Dream

This is a brief account of the theory put forward by Martin Seligman and Amy Yellen of the University of Pennsylvania in *Behaviour Research and Therapy* on ordinary dreaming (Seligman and Yellen 1987). They identified their broad form and the way the main parts are assembled. The essence of their proposal is that dreams consist of just three elements.

The first they call *visual episodes* or *visual bursts*. These are vivid images of brief duration whose essential feature is that their subject matter is random; it is free of links with anything that has just been dreamed beforehand, and uninfluenced by the emotional tone of the dream itself or by the subject's feelings before it started. They propose that their bursts can be identified by being vivid, central to the visual field, and capable of being scanned by the dreamer who can direct his attention to any part of them.

The second element which, together with the bursts, makes up the skeleton of the dream comprises the *emotional episodes* – loosely the dreamer's feelings. They say these probably have 'humoral underpinnings and humoral time courses', growing and declining over minutes or hours. These feelings have a physiological basis, such as real aches, thirst, discomforts; they may be caused by the topic of the dream; or they may reflect a dreamer's waking-life anxieties.

The third and last element is *integration*, which is the dreaming mind trying as well as it can to synthesize the emotional tone and the visual bursts into the least imperfect plot of which it is capable. The form of this integration is largely visual. It consists of a succession of images that are less vivid and less bizarre than the visual bursts, all of them self-supplied from the dreamer's own experience. By and large a dream consists of vivid random bursts – one of which is needed to start it – filled out with duller images that follow one another in a loosely connected stream.

Central to their whole theory is the proposition that the dreaming mind has to present a narrative whose development conforms to both the emotional tenor and the initiating visual burst. Thereafter each duller integrative image appears by involuntary association with its predecessor. Though tied more or less to the theme, this association is by a peculiarly imperfect process whose principle is conspicuously not by meaning as it is awake. The dreaming mind may juxtapose actual events or people or places that were experienced in separate unrelated waking contexts. Or it may select an image for its having aroused in real life the same feelings as those in the dream. Or the tie may be similarities in shape or colour or sound, for example the words like Gillian/Chilean associated by assonance rather than synonym. Seligman and Yellen call this feature 'adjacency', and it is enough to achieve coherence of a kind. They give this imagined dream as an illustration:

The bursts: Ronald Regan's face, a grapefruit, a rowing boat. The emotional tone: it's all very sad.

As a possible synthesis: Regan comes to speak in Berlin and we breakfast together; he has grapefruit which the waiter puts sugar on; Regan says how badly the summit is going, he might have to fire the Secretary of State; we are both sad at this and decide to go for a swim. Outside on the Wahnsee we see a rowing boat.

Seligman (who knew Berlin) and Yellen imply that the integrating images – Berlin, breakfast, sugar, the waiter, the

Secretary of State, swimming, and the Wahnsee – are all personal to an individual dreamer. Another person might link those three bursts in some quite different way.

For appreciable stretches of time, dream narratives develop gradually as the integrative images accumulate, e.g., a well-remembered house can by stages acquire new rooms and surroundings and inhabitants as the advancing dream-plot calls for them. Then a new burst forces an immediate and usually improbable change to the storyline, or it may start a new dream altogether if the discontinuity is too abrupt. Seligman and Yellen propose that the incorporation of external stimuli, like the dreamed sound of a door buzzer just as the bedside alarm clock goes off, depends upon the dreaming mind accommodating the intrusion in the same way that it does visual bursts. Both intrude into the dream story. They illustrate their principles at work in this second imaginary instance:

It is a grey and wintry day. There is snow on the ground. I seem to be at a four-way intersection, a crossroad, but I am not sure exactly where. Off to my right, at one arm of the crossroad, I see the back of a balding man, trudging up the hill away from me. I recognize him. He is an individual who has trespassed in my life and whom I hate. It occurs to me that I should physically attack him, and I begin to run after him. I stop and decide that physical violence would be foolish and wasted, and I notice that he is walking up towards a hospital anyway.

I now recognize where this crossroad is. It is located in suburban Philadelphia; to my back, the direction I am walking away from, is the house I lived in for seven years with my children and ex-wife. To my right up the hill is a hospital; to my left is a large Catholic monastery – beautiful, austere, and quiet. In front of me, the direction I am walking toward, is the University. I walk on a bit, in the direction of the University and see in front of me an undergraduate friend whom I have not seen in almost twenty years. I am very pleased to see him, and rush forward to embrace him, telling him how glad I am to see old friends once more.

Seligman and Yellen identify the balding man and the old friend as visual bursts. The crossroad that has significance for the dreamer is integration; it is synthesized, not clearly seen, and is a metaphorical four-choice crossroad in his life. The grey and joyless snowy day reflects his life's emotional low ebb.

The essence of their idea is that the content of the visual bursts has no explanation, the emotional tone is a background feature, and the integrating imagery is self-supplied from a personal repertoire of recollections by peculiar dreaming association. The authors emphasize that their theory is only about the structure of dreams. It accounts for the incongruity of dream events, for the style and transmutability of images and settings, and for the incorporation of external stimuli. It is an interesting and useful idea but I think there is one modification that could accommodate apparent short-term precognitions such as Maury's guillotine dream, one omission and one remaining uncertainty.

The modification is that visual bursts need not be assimilated immediately; though Seligman and Yellen say that one is needed to start a dream going that is not the chronology in the example just given. The grey day and the four-way intersection are both in place *before* the supposedly initiating first visual burst of the balding man. If the image of that man set the dream going then it must have been held off for a moment while his mind conjured up the snow and the crossroads to reflect his emotional state.

The Regan/grapefruit example did start with a burst – the president's face – but the other two bursts must also have had related synthesized images preceding them. The only point in breakfasting is to introduce the burst of a grapefruit; similarly there is no reason to swim but to introduce water for the coming burst of the boat to float on. Supposing the bursts in both dreams to be really random requires a mechanism that momentarily delays registration of them to allow time for dreaming integration to fashion a coherent link to them. If the image of the undergraduate friend in the second of their illustrative imaginary examples was in fact also a visual burst then by chance it appeared at a place in the dream narrative where incorporation posed no difficulty. The real dreams of the subjects sprayed with water are clear evidence for this same registration-delay.

Taking the delaying idea further, suppose a doorbell rings and the dreaming mind holds off the sensation while it fashions an appropriate link to the current story. If the ringing stops the dream might continue, if the noise persists it might awaken the sleeper – either way the dream narrative has the link in place.

The omission from Seligman and Yellen's theory is its largely discounting the presentation of features of a dream that are not imaged at all. In dreams I often just "know" something – that it is Monday, that this is Bahrain, that I am looking west, that the train is going to Manchester. These things one knows often appear early on in a dream by way of setting the background. It is the same idea as the film director who might over-write an opening townscape with the words "Paris 1947". In my own dreams much of the story is not presented as images, and the dream reports to come all show the same thing time and again.

The uncertainty in the theory is the provenance of visual bursts. The implication is that, although random, they are self-supplied like the integrating images; they say nothing about that and I think it may be significant.

Seligman and Yellen deliberately leave aside anything concerning dream physiology, or biological necessity or utility (psychiatric or otherwise), and they do not address any meaning dreams may have. Sticking strictly to structure their theory for the most part extends or expands earlier ones rather than significantly supplanting any of them. Their views fit well with those of the Marquis de Saint-Denys. He, in his own terms, distinguished between primary and secondary ideas in dreams, the former the main themes or sensory stimuli, the latter the dreaming mind's developments on them. He has 'transformation' when a new primary idea arises spontaneously. Saint-Denys' primary ideas are Seligman and Yellen's visual bursts and his secondary ideas their integrating imagery; his transformation is the change in the dream topic that follows the former interrupting a current dream sequence. The terminology is different, but each is discerning much the same thing.

Most discussions of dreams concentrate on 'long' ones with narratives, even though extended organized dreams are in my experience the exception not the rule. Often I get a mish-mash of indistinct, repetitive and dull images, and I have also noticed atypical features to dreams in shallow dozes around waking in the morning. The transition from sleeping to waking thought processes is fairly gradual and there is room for short intermediate states; going in the other direction – dropping off – I sometimes see rational thought beginning to fail gradually, irrationalities and non-sequiturs creeping into the stream of consciousness one by one without my mind being bothered to make any effort to correct them. I suppose that shallower sleep may be commoner in older people and more conducive to observing these states than the deep sleep of youth. All this is nothing more than my own experience, but I have no reason to think myself an exceptional dreamer. It seems to me that the long structured dreams upon which theories are founded may be minority cases. The assumption of a single set of principles derived from clear examples is fair enough, and useful, but it only relates to the structure of such dreams as have a comprehensive one. Some I would say do not.

Chapter 12

Features of Dreaming Thought Processes

Our minds are the conscious performance of our brains and that performance differs in characteristic and identifiable ways between waking and sleeping. In a sense we have two minds, a waking one and a dreaming one. Some of the differences between the two have been tied to activity at anatomical sites, but most of them are evident only by the products of that performance.

A particular difficulty with examining the imagery of any dream is that only a single individual ever knows what it actually was. It can be conveyed only very superficially to an experimenter. Other people's dreams, as well as our own unless they are very recent, are only accessible through notes made on waking, which are always incomplete. The brain site of memory being largely inactive during sleep may explain the imperfections.

We can see for ourselves that the differences between the sleeping and waking brain's performance are all deficiencies in the former. Absence of useable memory of everything we have learned of the world is the dreaming mind's most spectacular defect, and it brings an absolute lack of comprehension in its train. It attaches meaning to nothing, has no access to our past experience, no awareness of the way the world works, and no sense of normality – social, physical, or temporal. It resorts to a range of tricks and illusions to eke out this spectacular inability to understand, but that same inability allows it to get away with them. A positive usefulness of faulty sleeping memory is its not demanding of a dream narrative the slightest continuity. Impossible jumps in time or space do not seem absurd. The seemingly innate need to create a running narrative drives Seligman and Yellen's process of integration, but asleep our minds cannot distinguish what is likely and readily settle for the first absurdity.

Here are details of a dreaming mind's performance, most from my own experience. Much dream commentary concerns the outcomes of its workings rather than the process of arriving at them, but I dream every night and it did not take long to collect data; I have not the least reason to suppose that the material is in any way atypical. Almost everyone will match or surpass these examples if they have ever paid attention to the mechanisms of their own dreams, or soon would if they started now. It is useful to have an idea of what is going on but there is nothing authoritative in what follows.

Memory

At its simplest memory is the faculty that allows deliberate recall of the past, usually by meaning; and spontaneous association, usually by physical properties like colour, smell sound, behaviour, actions and movements, or spatial juxtapositions and contexts. Dreams lay down transient memory traces because on waking we can recall them after a fashion, but I have not found an earlier part of a dream recallable within it. Once I dreamed of seeing the diamond suit of cards, from the nine down but not in order; there were about two missing and I could see which they were. Then the image changed and I found I could not in the dream remember which cards had been missing, even though I had noted them only moments earlier. I could not recall them on waking either.

Reasoning

In 1987 the *American Social Scientist* Howard Margolis proposed that reasoning is a process of sequential pattern recognition. Recognising patterns is obviously a matter for memory, and would therefore, if Margolis is right, be debarred in dreams. I once dreamed of a long planning meeting about a manned

No Time and Nowhere

expedition to the Moon or to Mars. The planet was chosen because water could be had by drilling into the permafrost. An all-women crew had been mooted because men would have sexual problems, and we debated whether female drillers could be found. It all seemed perfectly rational in its own terms, even after waking but in fact it was just an illusion of reasoning; each point was referred only to what immediately preceded it and not to its actual context – e.g., drinking water is not the sole criterion for extra-terrestrial travel.

The psychologist Moers-Messmer reported in 1938 a lucid dreamer similarly unaware of context.

(*Walking on river ψ56*) He finds a river to have a firm but slightly yielding surface he could walk on, and many people were crossing it from bank to bank. He accepts it in his dream, but when he sees a bridge he starts to wonder what this material is: 'the weather is too warm for ice, perhaps it is a new invention, but if so why build bridges?'

– failing to make the obvious deduction that old bridges ante-date new inventions.

Knowing

We can look more closely at those things that a dreamer knows but does not image. There are sensations that seldom register in dreams, such as smell and taste. I have only ever read of once, and only once recorded one myself. I have dreamed of putting my hand under a tap and knowing that the water was tepid without sensation of wetness or warmth. In a dream in which rotten food featured I knew it stank but had no impression of the stench. The information that these sensations provide awake can sometimes contribute indirectly to a dream narrative.

Touch may be a little different; there is a range of tactile sensors – temperature, pressure, pain, stretching, tension, and more. The information they ordinarily give varies widely, but much of it is not imaged either. Most of it relates to our physiological ambience – pain, cold, wetness and so on; of that which is representative, the Braille alphabet, identifying coins in one's pocket by their shape, a blind man tapping along with his stick or running his fingertips over a face for example, all depend for imaging on the sensors moving. Because there is no such movement asleep that may impair tactile imaging, but it does not exclude it altogether. That cat which walked about on a bed at night is an instance.

A feature of knowing in a dream is that quite often one 'knows' wrong. Once I dreamed of a tall daunting woman dressed in a black suit with fair hair in a bun. Her dream identity was that of an aunt of mine who in life was a slight, colourful, lively, short-haired redhead. The dream woman simply *was* that aunt, despite the nonsensical absurdity of it. I take the fair woman to be illustrating Seligman and Yellen's visual bursts and integration, her appearance persisting despite the wrong identity but as the dream continued she spoke and moved in the way my aunt really did.

When we see in a dream an individual whose appearance and identity do not match it is defective recall or comprehension that prevents us from rejecting the absurdity, but there is something neither visual nor auditory that tells us who that person is, and the same holds for facts or objects. That identity is as much a part of the dream as any of the sensory-mode images, and is fully accommodated in our recollection and repetition of the dream story. The capacity to accept wrongly associated images and concepts compensate for deficiencies in sensation or under-standing, but even this positive feature is undermined by our inability to remember the waking world. If we know in a dream a fact that cannot be true we entirely fail to recognize its wrongness or ignore any interpretation that the dream ascribes to it.

Adjacency

Here are unusually clear instances of Yellen's 'adjacency' – association other than by meaning. It rarely happens awake so that when it does it provides useful confirmation of what is going on asleep.

In Singapore one night I dreamed of a poem called 'The Destruction of Tenasserim'. Waking moments later I recognized the poem's title as Byron's 'The Destruction of Sennacherib', and Tenasserim as the name of a place six or seven hundred miles up the Malay peninsula that had featured in a biography that I had read some fifteen years earlier of an Englishman in seventeenth-century Siam. The pronunciation matters because it was the similarity that led to the dreamed substitution – the 'ch' is hard as in *chronic* and the stress is on the second syllable, as with the Anglicisation of *Tenasserim* also.

Soon afterwards, an unconnected waking experience went like this. On my wondering, in the course of a conversation one evening, why songs so different as *The Red Flag* and *Maryland my Maryland* should share the same tune, the German wife of a business colleague said that the original was *Tannenbaum, O Tannenbaum*. The word meant nothing to me and I saw my mind dispatching runners to its archives to see if anything there might allow an intelligent guess. After an appreciable period, at least a couple of seconds, 'Tannenberg' came up – but the *sound* alone, without meaning attached. The runners sped off again to find out what the word meant and in about a third of the time the first search had taken there came back my personal association, which happened to be 'First World War battle in Solzenitzyn's *August 1914*' – no help in the circumstances. At the same moment the lady said "You know, Christmas Tree O Christmas Tree".

There are two things about this, the first the similarity of sounds in each – Tenasserim/Sennacherib and Tannenbaum/Tannenberg – both instances of association by assonance in the absence of meaning. Though common enough in dreams it is rare awake and it only arose here because, knowing no German, the word was to me literally meaningless and was not a symbol for an idea as words ordinarily are. My mind was denied its waking attribute of association by comprehension and instead had to make do with its next best option – its physical sound as it would in a dream – like those friends' names in the last chapter which were registered as like-sounding words.

The second thing is that my waking brain recognized that it had a problem and sought solutions to it but, denied comprehension, my dreaming brain could do nothing but settle for nonsense.

Contradictions

Dr Hearne was the first to notice that a dreamer cannot switch on an electric light; if he tries the bulb is broken or the fuse blown or there is a power cut (Hearne 1990). Dreaming minds seem unwilling to countenance situations that conflict with physical sensations, in this case without proper consequences like screwing up one's eyes. Here is an experience I had just before writing this. In my dream I was waking slowly to the realisation that indistinct noises downstairs meant there was an intruder in the house. The hall light from the top of the stairs did not work when I tried to switch it on before going down. I decided that light streaming through the open bedroom door behind me would be better than none, but when that failed to come on either I felt with a chill of apprehension that I faced a professional burglar who had turned off the power supply to the whole house. With that I awoke from a neat demonstration of Hearne's point.

In lucid dreams some subjects have devised tricks to make the light come on, such as looking away as the switch is pressed. Turning lights off in dreams is less difficult.

Restrictions on Content

Dreams indubitably incorporate actual bodily sensations but there is a clear hint that, forgetfulness notwithstanding, they will decline to accommodate too gross a discordance between what is sensed and what is being dreamed. Because muscular awareness and motor control are blocked while sleeping there can never be any contradiction to dreamed movement and, eyes being shut, the same goes for seeing. A striking laboratory discovery is that seeing and moving comprise the bulk of all dream material. Speaking does happen, sleep talking is an extreme case, but researchers say that a dreamer is more often spoken to or planning what to say than actually speaking himself, and I see that repeatedly. Usually the sense of what is said is known without any visual or auditory images of words.

Reading

Most dreamers report that reading is impossible, and some use this as a test for lucid dreaming. Once I was able to discern nothing but a jumble of 'k's' and 'x's' in lines of dreamed print; another time I was watching words of my own dictation being transcribed into Times Roman capitals, imaging each word in turn and sometimes two or three together, but longer sequences turned fuzzy and were forgotten. Everyone must have a vast repertoire of images of printed words available to illustrate dreams just as other stored images do, but extracting meaning from them cannot be done. What usually happens to my closely observed dreamed text is that it keeps changing, or fades, or melts away like a Dali watch, but sometimes it leaves an impression that it *has* been read and its sense conveyed. May be this is a refusal to countenance the contradiction of being unable to read.

Calculation

The impossibility of calculation in dreams is self-evident. Once I was unable to multiply two digits by a single one, a sensation of utter helplessness. That dream was only remembered following another soon after which contrasted with it. In the second, I could get the moulded depth of a ship by adding plus 17 and minus 14 = 31 (they were measurements above and below the waterline, hence the signs). Adding is a simpler process than multiplication and may perhaps be automatic if the numbers are small. Green and McCreery have a case of a lucid dreamer unable to divide 200 by 5.

Amalgamation. I once dreamed of finding two objects while tidying our house for a party – first a painted toy car and then a white plastic hand-held pencil sharpener with the design of the Finnish flag on it. Each image arose distinctly and separately, but they merged to yield a third image of a white car with the blue cross on its side. On another occasion a particular chisel and a screwdriver that I own merged to form a composite image of a single tool, quite automatically and utterly pointlessly, my dreaming mind just did it. Amalgamation in dreams is not uncommon, but unless particularly noted it gets subsumed into a logically more coherent narrative and is consigned to waking memory (and any record) as the combined entity. Amalgamation can dispose of awkward discordant juxtapositions, or simply be economical.

Interpretation

Of similar effect is arbitrarily re-interpreting images or external stimuli to render them less exceptionable to the current narrative, and taking the necessary time to do it. Dozing lightly one morning I was dreaming of a spacious country house drawing room at night with dark old-fashioned curtains and furnishings, lit as though with candles. Suddenly there was a flare of bright white light, taken by dreamed knowledge to be someone striking a match, though the flare was more like a firework, too fierce to have been hand-held. On waking at that moment I opened my eyes to find the sun shining on to my face through a chink in the curtains with the colour and brightness of the dreamed light. (The optical input must have been red light through closed lids; my imaging process adjusted the perception to real-world expectations, and created the flare in the process.) Besides a sunbeam, that dreamed room could have accommodated almost any sensation impinging from an urban environment, a police siren warning of Bolsheviks in the grounds, a motorcycle back-firing the young master shooting pheasants, a smell of burning the ambassador's cigar, music the village fair, and so on. There is nothing that could not have been plausibly introduced and a dream does not even have to be plausible. Re-interpreting real sensations in dreams happens of itself often enough. It is no indication of short-term precognition.

In Maury's case quoted earlier he was already dreaming about the Reign of Terror, so that incorporating a blow on the back of his neck as a falling guillotine blade was not far-fetched. Had he been dreaming of anything else some other integration would have been called for, say a broken tile falling from a building as he passed.

Discarding

Yet another mechanism for simplifying dreams is the dumping of inappropriate images that disrupt the storyline, simply discarding them and forgetting that they arose. It is obviously difficult to spot and I have only ever noticed it once. In a macho dream, I was one of a party escaping from a prisoner-of-war camp to carry out some unspecified military activity ten miles south-westwards. We were crawling about under huts and snipping wire, and then tried to bluff our way past a manned gatehouse. In the middle of all this there appeared an image of a friend in Dublin. She was just there, playing no part in the story and not influencing it in any way.

The dream arose in the middle of the night. I had nothing by the bed to write with and did not feel like getting up. But I kept thinking about it, and went downstairs to record it some minutes later. Only on reading through the account before going back to bed did I notice that the friend had been left out, forgotten already. There was nothing striking about her image. It was related to nothing that preceded it and was not incorporated in the scene, but it did for a moment occupy the whole visual field and it was certainly random. I have no doubt that it was a visual burst, but it did not start a new dream nor force the ongoing narrative into a new direction so there must be occasional exceptions to Seligman and Yellen's proposals on that score, and to Saint Denys's too.

Creativity

Finally, I sometimes wake up with yesterday's ideas more clearly arranged in my mind than they had been, or to a felicitously precise way of expressing them, an experience which anecdotally seems common enough. Much rarer and more striking instances include these: Coleridge claimed to have dreamed his poem *Kubla Khan*; Kerkule is supposed to have dreamed the structure of the benzene ring; there is a case, one of Green and McCreery's, of a man who during the Second World War was having difficulty repairing an old mechanical adding machine and then dreaming correctly how it should be done; the Slovenian musician, Ljuben Dimkaroski, dreamed how to play a replica of a fifty-five-thousand-year-old Neanderthal flute that neither he nor anyone else had managed to play before. In none of these remarkable instances was the result accessible to logical reasoning because there was insufficient data to reason from; the dreamers' minds just jumped.

* * *

Any dreaming mind has to accommodate intrusions from both visual bursts and external stimuli into as unexceptionable a narrative as it can. Freud saw the preservation of sleep as important and dreaming seems essential to our health; both are achieved by eliminating whatever disturbs it. Besides motor paralysis there seems to be a mechanism for achieving peacefulness through the mind's modifying, amalgamating or ignoring some images, and taking some of them for what they are not. The dreaming mind has no facility for relevant recollection, nor ability to read, reason, or calculate, to comprehend facts or to recognize anything at all – faces, symbols, absurdities, discontinuities, errors. Dreaming is a distinct and peculiar state of consciousness. A dream is no more than the immediate spectacle of a discontinuous stream of meaninglessly associated images unfolding scene by scene to illustrate very patchy information. It has no causal coherence, no logic and no predictable development.

The informational content of any dream is low by waking standards. If we awaken from a dream with information on something that happens later , if information seems to have been conveyed to us in some way as facts or images, we have understood it in extraordinarily imperfect ways. Dunne stressed the importance of recording dream images themselves as well as the interpretations put on them. Ideas of significance, if any, are indistinguishable from all the rest. No dreamed precognition can be judged as if it were the equivalent of a recollected waking event, and elements that may seem to have been precognitive have to be evaluated in that light.

The workings of dreaming minds throw a somewhat different light on recorded precognitions. Discrepancies with later events have no relevance whenever they are artefacts of the medium, and the correspondences, the hits, loom a little larger in consequence.

Chapter 13

The Content of Precognitive Dreams

Precognitive dreams vary widely in their forms and presentations. One way to categorize them is by content, the first being those experiences which comprise foreknowledge of the arrival of news on an event that has already occurred somewhere else. They differ from premonitions of danger by arriving asleep and being registered by a sleeping mind.

Foreknowledge of News to Arrive

Here are three from Dunne: At Alassio in Italy in 1901, having been invalided home from the Boer War he dreamed:

> *(Fashoda: ψ57)* ... that he was in Fashoda, a town in the Sudan where Anglo-French colonial competition had caused a 'diplomatic incident' a few years previously, when amongst other things going on, three men appeared from the south. They were sun-burned almost black and wearing faded ragged clothes rather like the troops he had just left and he wondered why they had journeyed from the south to the Sudan and asked if that was so; they replied that was exactly what they had done, 'we have come right through from the Cape' (of Good Hope).

Next day's *Daily Telegraph* – which had taken several days to reach him – carried the arrival of its trans-Africa expedition in Khartoum (African exploration was big news at that time). Dunne had heard some years earlier that the expedition leader, a Mr Decle, was contemplating such a venture but he did not know that anything had come of it. Since it was an old paper, though new to him, he could not suppose that some kind of contemporaneous clairvoyance was at work, and he attempted no explanation.

The next dream has often been quoted: In 1902 he was back in South Africa in camp when he had an unusually vivid and disturbing dream which has been quoted often. He dreamt that:

> (*Mont Pelée* ψ58) ... he was on high sloping ground of a curious white formation with little fissures in it from which jets of vapour were spouting into the air; he recognized the scene as one that he had dreamed of before – an island in imminent peril from a volcano – and realized in distress that it was about to explode; (he had read of Krakatau, the island volcano in Indonesia that utterly disintegrated in 1883); in his dream he was seized with a frantic desire to save the four thousand inhabitants, which could only be done by sea; he found himself at a neighbouring French island trying to get incredulous officials to despatch every vessel they had, and the dream became a nightmare as he was repeatedly told to return in office hours; he awoke clinging to the heads of the carriage horses of 'Monsieur le Maire' who was going out to dine, shouting to him that four thousand people would die.

When the *Daily Telegraph* arrived it reported the eruption of Mont Pelée and the destruction of the town of St Pierre on the French island of Martinique; there were various items concerning ships. The report spoke of 40,000 lives lost but Dunne misread this as 4000, and re-told it so until copying the text from the paper fifteen years' later. (The error turns out to have some interest.)

News of the eruption travelled from Martinique to London, no doubt in half a day or so by telegraph, and thence in two or three weeks by sea and rail to Dunne. The dreamed event was neither the eruption nor the printing of the paper but, as for 'Fashoda', receipt of the news.

Dunne's Mont Pelée dream had suggested the unpleasant

possibility that he might be suffering from what he called 'paramnesia' – by which he meant wrongly imagining after the event that he had actually dreamed of it beforehand – so he recorded every detail of this one as soon as he awoke. He dreamed one night:

> (*Factory Fire: ψ59*) ... that he was standing on a high wooden walkway with a deep gulf beyond it filled with thick fog limiting visibility to three or four yards:
>
> ...then he noticed a thin lath-like thing projecting up from the gulf, slanting inwards towards the awning; it began to wave up and down brushing against the railing; and after a little while, in his dream, he realized that it was the jet of a fire hose, a sight that had previously puzzled him in a film about a fire; the dream then turned to horror; the walkway became crowded with people who were gasping and collapsing as the smoke grew thicker and blacker until his only sensation was of the moaning and suffocating people.

The evening papers brought the news he had been half expecting. There had been a big fire the previous day near Paris in a factory that produced some rubber-like substance which burned noxiously. A large number of girls had been cut off by the blaze and had made their way on to a balcony in the open air. The fire brigade's ladders were too short to get them down, so they sent for longer ones and sprayed water on to the balcony in the meantime. But before the long ladders could arrive the windows behind the balcony broke, and the smoke released was so dense that every one of the girls suffocated. The elements relating to the event being a fire were suggested by limited visibility, by many people dying of smoke inhalation, and by the hose jets. Height was suggested by the railing and a deep gulf.

Dreamed Precognitions of Actual Events

The next category comprises those dreams ordinarily thought of as precognitions for concerning events that have yet to happen. They can be further broken down into those that relate to public or private calamities that occasion a negative emotional response by the dreamer – anguish, fear, distress; and those whose content is purely trivial.

The first of the disturbing category is this famous dream from the late nineteenth century that I came across in J.A. Hadfield's *Dreams and Nightmares* (Hadfield, 1954). Its original source is the Proceedings of the Society for Psychical Research. A Lady Q lived with her uncle, a retired colonel who was as a father to her. One night she dreamed:

Lady Q's Uncle: ψ60) ... that she and her sister were sitting in the drawing room of his house on a bright spring day with a dusting of snow over the flowers in the garden. In the dream she knew that her uncle who had been riding in a dark homespun suit had been found dead in a certain spot, his horse standing nearby; she also knew that the body was being brought home in a two-horse wagon with hay on the bottom; then she dreamed of its arrival at the house and of two known men carrying the body upstairs with difficulty on account of her uncle's large size; in the process the body's left hand hung down and struck against the banisters as the men ascended – a detail she found so horrible that it woke her.

In the morning she made her uncle promise not to ride that way again without a groom in company. Two years' later Lady Q had the same dream again, her uncle when taxed admitting that he had occasionally broken his promise. Four years after that, having moved away, she was lying ill when she had a repeat of the dream in all but two particulars; she now saw herself in the bedroom of her house in London, and a man in black whose face she could not see came in to say that her uncle was dead. Lady Q wrote to her uncle about it. He died two days' later but his death was kept from Lady Q during her convalescence. When her father-in-law entered her room dressed in black to break the news, she cried out 'the colonel is dead. I know all about it. I have dreamed of it often.' Except for the snow excepted, events at his house had matched the details of the dreams in all recorded particulars – the two men who carried the body were the two she dreamed of and the left hand did strike against the banisters.

This is another well known experience from Dunne. In 1904 he was holidaying with his brother in Austria when he dreamed that:

(Escaped Horse ψ 61) ... he was walking between two large park-like fields down a sort of pathway which was flanked on either side by high iron railings; in the left-hand field a huge black horse was galloping about kicking and plunging in a frenzied way, and Dunne glanced along the fence to make sure it could not get out. He saw no way for it to do so, but just then galloping hooves sounded behind him and he saw that the horse *had* got out somehow and it was coming after him; he fled towards a flight of wooden steps ahead of him, and awoke just before he reached them.

Out fishing next day his brother exclaimed at the behaviour of a horse on the other side of the river. As in the dream, Dunne saw the path between the two fields, the agitated horse, and the wooden steps which led up to a bridge across the river. The general scene was right but the details were entirely wrong – the fences small and wooden, the fields small too, the horse of ordinary size and to the right of a man walking towards the steps rather than on the left. He started telling his dream, remarking (after checking the fences) 'At any rate this horse can't get out'.

But almost immediately it too got out somehow, no doubt it jumped, and was tearing down the path to the steps; there it swerved aside and charged through the river towards the two men, who picked up stones and retreated from the bank. As the horse got out of the water it stopped, snorted, and then galloped away down the road.

Dunne wrote (I have taken slight liberties with his syntax):

... one thing was abundantly clear. These dreams were not *percepts* (impressions) of distant future events. They were the usual commonplace dreams composed of distorted *images* of waking experience, built together in the usual half-senseless fashion peculiar to dreams. That is to say, *if they had happened on the nights after the corresponding events,* they would have exhibited nothing in the smallest degree unusual. Regarding the waking experiences which had given rise to them, they would have yielded just as much accurate information as any ordinary dream does – which is very little. They were the ordinary, appropriate, expectable dreams; but they were occurring on the *wrong nights.* (His italics)

After this dream Dunne concluded that he was experiencing a displacement of time in some way. Though quite inexplicable, he was pleased to have replaced separate suppositions for each odd dream with a single one for all of them.

Dame Edith Lyttleton reports that a Mr Lloyd dreamed:

(R101 ψ62) ... of a large airship crashing after manoeuvring difficulties. He saw it hit a hill top and burst into flames. He saw people failing to escape the flames, a small company of soldiers, and a mounted officer dashing about.

He told his fiancée about it. Three days later on 6th October 1930 the R101 airship drove into a hill top near Beauvais in France, broke up and caught fire with numerous casualties. A mounted French police officer was prominent in the cinema newsreels that Mr Lloyd saw. Dame Edith was told of another person who dreamed of the R101 as well. This dream occurred about six months before the event and it was repeated one week later.

Rhine presents this case:

(Typewriter Customer ψ63) Two couples in California had adjacent businesses, Mr and Mrs A sold sewing machines, Mr and Mrs B repaired typewriters. One customer for the typewriter shop found it empty; Mrs A, who had met him about a dozen times, invited him to wait in her shop. When Mrs B arrived and was talking to her customer about his needs Mrs A turned towards him 'with a look of terror on her face' so that he asked if she had seen a ghost. She rather thoughtlessly replied that she had dreamed of him the night before, and then had to tell the young man that 'she had seen him and his wife driving through an underpass and running into a truck. His wife was hurt and he decapitated'. A few weeks later that is just what happened.

This one is from Priestley's collection. In 1938 a twenty-year-old man had a nightmare which started with:

(Grenade Injury ψ64) ... an explosion and his becoming aware of a stabbing pain in his left thigh, right forearm, and left temple and eye; in the instant before he awoke he saw 'the blood as it left my wounds'.

In 1940 at Dunkirk he was wounded by a grenade thrown into a small country inn. He was wounded in his left thigh, right forearm, and by a splinter near his left eye that filled with blood. He went to a mirror to test the sight of his eye, wiping it with a handkerchief and shutting his right eye.

Dr Hearne recorded this from the 1980s: sixth-form schoolgirl dreamed that:

> (*Sailing Accident* ψ65) ... she and the younger brother of a classmate were crossing a nearby lough in a sailing boat which capsized; in the dream she got herself clear of ropes that entangled her but was unable to get the boy out of the same difficulties.

Next day the girl told the dream to a female lecturer who was a colleague of the boy's father. It was she who phoned the pupil that evening to say that the boy, who was a good helmsman, had been drowned when his boat capsized in bad weather on the lough that day. Hearne includes a very similar case concerning a boy accidentally shooting himself dead while wild-fowling.

Hearne also reports a woman waking her husband in panic after dreaming:

> (*Tenerife Air Crash* ψ66) ... that two massive aircraft had crashed, but not from the sky; she was standing underneath one of them as passengers screamed to her for help. She knew that the incident was on the ground and at Tenerife airport.

She told several friends about it over the ensuing seven days. On 27th March, 1977 in thick fog at Tenerife a KLM 747 taking off collided on the ground with a taxi-ing Pan Am 747; it still stands as the worst aviation accident, nearly 550 people were killed. Another woman told Hearne that she had dreamed the night before the accident of a horrific head-on collision between two planes, one landing the other taking off.

Finally there was foreknowledge of the 9/11 attacks. Soon after, Dr Sheldrake appealed through local advertisements for any intimations of the attacks and received more than fifty replies. Among them given in *The Sense of Being Stared At*

(Sheldrake, 2003) are these three:

(Twin Towers ψ67) Five nights beforehand a New Yorker dreamed that he and his fellow passengers were all alarmed by the plane flying so low over the city; he saw they were over the southern end of Manhattan; then there was a 'tremendous impact' and he awoke.

A woman dreamed the day before that she was in a plane with a long-haired dark-skinned man on the opposite side of the aisle; she could see the cockpit instruments and through the windscreen a building which the plane hit; both plane and building caught fire and then the building began to fall.

Another woman dreamed in the early morning of the day that she was in the World Trade Center 1; when it caught fire she escaped across a bridge about mid-way up into a second building which then caught fire itself.

Precognitions of Trivia

An easily distinguishable class of precognitive dreams are those whose content is utterly trivial and the dreamers emotionally indifferent to it – except perhaps for surprise.

Priestley found that of the apparently precognitive dreams sent to him about forty-five per cent dealt with calamities, like Mont Pelée, the Factory Fire, the R101, the eruption on Heimay, and the Twin Towers. Another forty-five per cent of them concerned matters of no importance whatsoever. He found their stark inconsequence more persuasive than awesome calamities just because they had no conceivable prompt in the dreamers' lives and had no practical significance. All are very improbable happenings. Among them are these, the first four from Dunne and then three from Priestley: A Mrs Atlay, the wife of the Bishop of Hereford, dreamed that:

(Bishop's Pig ψ68) ... because the Bishop was away from home

she had read family prayers, not in the chapel as the Bishop did but in the hall, and that on completion she had gone into the dining room to find an enormous pig standing between the table and the sideboard.

The Bishop really was away. As she, her children and their governess waited in the hall for the servants to come in for prayers she amused them all by telling her dream. After prayers she opened the dining room door to see 'the pig in the very spot in which I had seen him in my dream'. Its sty door had been left open during cleaning and the door from the dining room to the garden was open as well. This account was first written in 1893 but the event had occurred some years before. The governess vouched for it.

At one point Dunne wondered whether he could capture future impressions by allowing his mind to drift aimlessly without actually going to sleep. He tried it with novels, holding them as he sat in a deep club chair searching about for intimations of their contents. When that produced nothing he tried without a book at all, but only once did anything come of it.

There came into his mind:

(Umbrella ψ69) ... the picture of a closed umbrella standing unsupported upside down on the ground outside the Piccadilly Hotel in London. Its handle was a straight slightly thicker continuation of its shaft, with no bend to it.

It sounds like a modern golfing umbrella. The next day as he passed along Piccadilly on a bus he saw walking along the pavement an old lady eccentrically dressed in early Victorian fashion. She carried a straight umbrella like the one he had seen. This she was using as a walking stick, pilgrim-fashion, but she held it upside down, the ferrule end in her hand, the handle on the pavement. Dunne was quite sure he that had never before seen an umbrella being used like that.

Dunne went on to recruit friends and relatives to record their dreams for him. His cousin Miss B at first could remember no dreams at all, but she persevered. This was her sixth dream:

> (3 Cows: ψ70) ...she was walking along a known path and found at the end of it a five- or six-barred gate that had no business to be there; beyond the gate a man passed driving three brown cows, holding a stick out over the animals in an odd way, as if it were a fishing rod.

Later that morning she walked to the end of the platform while waiting for a train at Plymouth station. There was the gate of her dream with the man, cows, and stick, 'all arranged exactly as in the dream'.

One of the consequences of the publication of Dunne's book was a play called *Time and the Conways* by J.B. Priestly and another John Buchan's book *The Dreams of Four Men* about a quartet who between them dreamed a page of *The Times* one year ahead. There was also a three-week trial by members of the Society for Psychical Research conducted under the auspices of a Mr Besterman in 1933. When nothing notable came of it Dunne suggested that younger subjects might do better, at which the Society organized a trial among students at Oxford. Only two of them completed the full series of three weeks' observations with notes made on waking. One of them dreamed:

> (Cartwheel Boat ψ71) ... I was sailing alone in a small boat constructed out of a number of cartwheels cut in half with planks nailed along and tarred.

He appended a rough sketch; the craft was the shape of a half cylinder cut along its axis of rotation; there was a half cartwheel at each end and one in the middle, the planks being fixed to their rims. Nearly a month later he wrote to say he had seen an object 'precisely the same as that seen in my dream'; he was in the company of a vicar at the time who confirmed that a man had been carrying "a curious object of the shape drawn below" up the hill outside the vicarage "(3 half circles joined with slats of wood)". (There was a rough sketch of an object very like the first one except that the spokes were missing.)

These are among Priestley's cases: A man wrote to say that when he was a student he dreamed that:

(*Sparrowhawk* ψ72) ... a sparrow-hawk was perched on my right shoulder; I felt its claws.

Sometime later he was studying in his shared digs when the landlord offered them for their fire wooden rubbish he had just cleared from the attic. One of the items was a stuffed sparrow-hawk mounted on a board that had once had a glass cover. The young man paid no attention and continued working, but one of his friends detached the bird from its base, crept up behind him and clamped it on his shoulder firmly enough for it to stand unsupported. He felt its claws and that reminded him of his dream.

A woman told her friends at breakfast of dreaming that:

(*Thirty-three eggs* ψ73) ... a farmer arrived with thirty-three eggs in a bucket, and a little later three more eggs were handed to her while half way up the stairs.

Shortly after breakfast was over a farmer handed her a bucket of eggs saying it held three dozen. They were leaving that morning so her husband took them upstairs to pack them, soon after calling down that her dream was true because there were only thirty-three eggs and asking her to come up and check them. As she did so she was called down to meet on the stairs a woman who said there had been a mistake and gave her another three eggs.

The director of an amateur dramatic group in West London announced that their next production would be a still-undecided play by Pirandello. A young woman who had never heard of that playwright dreamed that night:

> (*Pirandello's Bat ψ74*) ... of being in Italy (which she had never visited) in a long narrow room with arches along one side leading to a rose garden. She was one of a group of medieval people dining at a long table. Her partner led her into the garden through one of the arches, and as they passed through a bat flew in from the garden; she reached up and caught it.

Next day she looked up Pirandello in the library and 'the very first play I started to read had as its setting the exact room of my dream complete with the rose garden seen through arches. More amazing still, later in the action was the most improbable stage direction I have ever met – "a bat flies in"'.

Dreaming Horse Race Results

This next set of experiences is also trivial, though exciting too. Racing gives a useful slant on precognition because the facts are simple and the outcomes unarguable. I came across this extraordinary series of dreamed horse-race winners in a book called *Tell Me the Next One* by John Godley (Godley, 1950). He was a cousin of the Miss Godley who saw Robert Bowes on the lake in Chapter One.

After distinguished service in the Second World War Godley returned to his studies at Oxford and graduated in 1948; in March of 1946 he dreamed the first of a series of six precognitive racing successes that extended over three years. He mentioned the first three to friends or family in advance of the race, and bet on each one. Thereafter he wrote them down as well. It all excited a good deal of interest in his own circle, and he hit on the idea of selling the story to the press. The upshot was, at some remove, his abandoning the Foreign Office as a career in favour of journalism, and he became a racing correspondent for the *Daily Mirror*.

He had a punter's interest in racing initially and a student's perpetual need for money but, as one can imagine, his interest increased greatly as more precognitions came in. He never dreamed correctly the name of any horse that he did not know already. The others were approximations, e.g., "Tubermore" for *Tuberose* or "What Man" for *Mr What*, but, since most had another detail or two as well, all were close enough to be unambiguous. Two of the dreams featured two winners, so he got eight altogether on six occasions. If the facts look a little thin bear in mind that each dream carried with it a conviction that it referred to a win in an imminent race. In brief, here they are: *(Godley's racing dreams ψ75)*

Dreamed of a newspaper report that *Bindal* and *Juladin* had won at Plumpton and Wetherby respectively; Godley bet on both and that afternoon each did.

Dreamed of reading that *Tubermore* had won an unspecified race; he saw that *Tuberose* was running in the Grand National, he and his family bet heavily; *Tuberose* won.

Dreamed a telephone call in which his bookmaker said *Monumentor* had won at 5-4; he lodged a sealed letter with its name in the Post Office, told friends and bet; that day *Mentores* won at 6-4.

Dreamed of seeing a particular jockey winning in the Gwaekar of Baroda's colours, and hearing that *The Bogie* had also won; the same day at Lingfield that jockey won for that owner and

The Brogue won the next race.

Now a journalist, Godley dreamed a list of winners in a newspaper office showing *Timocrat* winning at Cheltenham; that day it did.

Dreamed a list showing *Monk's Mistake* winning at Taunton and *The Pretence* at Lingfield; he bet enough for the winnings to allow him to live without working for two years; only *The Pretence* won.

The National winner came in at 100-6 but otherwise, barring the last, an 80-1 double of which only one horse came in, the odds were all less than 4-1. There was one complete failure where the horse was unplaced. *Monk's Mistake's* jockey Edgar Britt was a friend of Godley's and told him the story of the race: three fences from home his mount lay ten lengths in front of the field and three behind the leader who was fading; Britt was certain he had the race by six lengths or more. Both horses rose as one for the last jump, but *Monk's Mistake* slipped or misjudged it. He just clipped the top of the flight, stumbled on landing and was unable to recover in time. The horse had started at 10-1 and *The Pretence* at 8-1; on the double Godley stood to win £2000. He wrote his book soon after that race and it generated a large mailbag from people after easy money, whence the title of his book. In a telephone call in October 2000 to Lord Kilbracken (as John Godley had become) he said that all the dreams were spontaneous. Though he 'tried' on occasions, by memorising all the next day's runners and having a pencil and notebook by the bed, nothing came of his preparations. When I spoke to him he told me he only ever had one more such dream, seven years after his book came out. He was in Monte Carlo and dreamed three months before the event that *What Man* won the Grand National. He won £450, worth all of £5000 in today's money; the details are given in the next chapter. He never had any other precognitions in his life, by dream or otherwise, before or after, racing or not, but his proof is the hardest you can get – cash. To Godley, all these dreams, including the part and total failures, seemed to have a distinctive quality.

I have heard it said that Godley was a grand raconteur not above favouring effect over exactitude, and that when he wrote his book he had just inherited his father's estate in Ireland which was in a bad way financially. That may be so, but the book reads matter-of-factly and there are plenty of unrealized opportunities for embellishment. Our telephone conversation and his following letter were also quite ordinary.

Content and Emotion

There are two other distinctions between trivialities and catastrophes besides their subject matter. In trivialities the percipient is present in person at the event but not at the catastrophes. And trivialities have no emotion associated with them while a catastrophe is distressing just to hear about, the emotional response it evokes is strong and negative. This negative emotion attending calamities matches that in crisis cases; they too are confined to distressing circumstances except for the rare instances where they are visualized – Lord Brougham, the First World War airman's sister in India, the figure of the commissionaire called 'the Major' in Chapter Three. We live in rational times and as a general rule seek where we can to exclude feelings from our thinking, men especially perhaps, but the central position of emotion in two distinct classes of non-sensory information – calamities and crisis intimations – may suggest that it is ordinarily central to some intellectual activities.

Chapter 14

Provenance of Imagery; replications, quantity of information; latency

We all can recall images of past scenes and of people we have met, and we all have, perhaps to differing degrees, a creative imagination that can juxtapose past images or parts of several of them to produce something new. Are the images which illustrate precognitive dreams of the same kind or are they future scenes to be experienced later?

Dreams without images

Many dreams include elements that are not imaged at all. How did Dunne know he was in *Fashoda*, or that the men had come from the south? *Mont Pelée*, how did he know the ground was high? He supposed an awning *Factory Fire* dream but could not see it. *Lady Q* – how did she know her uncle was dead and was being brought home, she reports no images of that before dreaming of his arrival? There are two more instances of the same thing later in this chapter: Dunne knew he had invented the flying canoe, and the *Titanic* dream was meaningless without the ship's name but neither fact was conveyed by any images. Dr Hearne's reports are mostly too short to distinguish an image from a received idea but in the Tenerife dream the centrally important fact of the location was not imaged. Much dream material is received as facts alone.

Images of apparently future provenance

Some imagery seems to be drawn from the real scene experienced later because it concerns events that are so very unlikely to have been seen before. These are examples: the three cows with the stick held over them, the upside down umbrella, the Coptic rose, the sparrow-hawk, the umbrella, the unknown horse names, which come later in this chapter, and perhaps the cartwheel boat. In this next dream the imagery seems to have been future-based in order to permit visual recognition when Mrs Schweitzer met Deverell later. She dreamed in 1882:

> (Mrs Schweitzer ψ76) ...that she saw her younger son on some cliffs with a stranger when the son suddenly slipped and fell. She asked the stranger his name and he replied 'Henry Irvin'; she then asked 'Do you mean Irving the actor' and he answered 'No, not exactly, but something after that style.'

Next morning she asked her elder son to recall his brother from his business trip to Manchester, but he took her fears lightly. Eight days' later on the 26th of July the younger son was on holiday in Scarborough when he fell from a horse and died within a few hours. The day after Mrs Schweitzer reached Scarborough, she went to view the scene of the accident accompanied by her son's companion at the time of his fall, he sitting opposite her in the carriage. She wrote 'when I looked at him I remembered my dream of the 18th and recognized the stranger' (from it). She asked if his name was Henry, and when he said it was she repeated her dream. The man used to recite at private theatricals, where he was introduced as "Henry Irvin Junior". His real name was Deverell, and he had first met the younger Schweitzer only on the morning of the ride.

A woman in Ireland reported this uniquely extraordinary experience to Priestley.

She dreamed she was driving on a road she knew well when suddenly there was a little girl with dark curls and a bright blue cardigan right in front of her car. She hit the child and, to her horror, killed her. Next day she drove along that stretch on the way to her daughter's house and reported that as she did:

(*Blue cardigan* ψ77) "On approaching the spot, I looked round most carefully for any sign of children, and there were none, only about five people standing at the bus stop. Relieved beyond words, I glanced down at my speedometer to check, and on lifting my eyes, was completely horrified to see, standing still in the middle of the road, the little girl of my dream".

The woman slowed down and stopped just alongside the child, who just stared at her. There was a bus stop nearby and the women in the queue seemed more interested in the car stopping than to any danger of a small child alone on a busy road. As she drove slowly away she could see in her mirror the girl still immobile in the middle of the road. Her daughter was much relieved to see her, saying:

...last night I had a terribly vivid dream... you ran over and killed a lovely little girl dressed in a bright blue cardigan with lovely dark curly hair.

Priestly writes that the professional storyteller in him felt suspicious of the neat 'twist' and 'pay-off' at the end which seemed too good to be true, whence his writing for confirmation from the daughter and the woman's husband.

The driving woman's behaviour was odd for driving off and leaving the child in danger. This experience is unique and any classification of it therefore provisional, but if one takes the waking hallucination as the foreknown event its imagery indubitably derived from the woman's personal future.

Images from Past Experience

Imagine you are telling a friend the plot of a television programme that you had watched the night before. Suppose you start by saying "There was this American who drove to work from the suburbs each day" – as you speak there comes into your mind image after image in as much detail as you care to put into recalling them: his figure, his dress, his family, their house, its garden, the pets, and so on, with all your associations and assumptions as to his wealth and social standing; and then his drive to work, the neighbourhood, his car, and their associations also; his office, its business, number of employees, his position and status at work. But in the mind of your listener there is only the merest shadow of all this, the barest minimum needed to visualize from his own store of images the dozen words you have used. The informational content is externally supplied by you but the imagery is entirely his own. Had that listener re-told the plot to someone else he would have repeated your words, or described his own visualisations, or combined the two. If a story required an image of a tall bearded redhead with an eye-patch and a peg leg and the reader or listener had never seen such a sight he could easily assemble a composite image from the details he has been given. Anyone reading that sentence may have just done so. Visualisations of symbolic inputs to the human mind are most often the work of the recipient not the donor.

Foreknowledge of an event which is not witnessed personally can only be experienced as news of it. Nowadays news is usually illustrated but in earlier times it could only be imaged by the percipient himself – as with Dunne's Fashoda and Mont Pelee, and also this one. An Englishman due to travel to America in 1912 on the maiden voyage of that ship dreamed that:

(Titanic: $\psi78$) ... he saw it keel upwards with passengers and crew swimming around; next night he dreamed the same and told various friends about it.

A few days before sailing he received a cable from America suggesting for quite ordinary reasons that he postpone his journey. The imagery was his own and bore no relation to the reality at all.

The ubiquity of television does not exclude personal imagery, but the *Twin Towers* dreams were not illustrated with real images either, even though the televised pictures were very dramatic. All ordinary dreams are imaged by self-supply and it seems as though the same is true for most precognitions as well, with the most notable exception of trivialities.

* * *

Replications

Two of the preceding reports mention another precognition of the same event – the R101 and Tenerife, and the examples given for the Twin Towers were three of fifty. The report that follows shows a single central idea being presented in a number of different ways, and demonstrates that the effect is sometimes widespread. Aspects of a single event were presented in imagery personal to each percipient; they include four waking experiences as well, left there to keep the material together. This is the single event:

> (*Aberfan:* $\psi79$) At nine-fifteen one morning on 21st October, 1966 there occurred a horrific accident in the mining village of Aberfan in South Wales. The coalmine's massive spoil heap towering above the houses slumped after a period of heavy rain. It swallowed up the part of the village that included the school – of 144 people killed, 128 were children.

John Barker, a psychiatrist who became involved in the aftermath of the tragedy wondered whether there might have been foreknowledge of so starkly distressing an event. Through a journalist friend on the London *Evening Standard* he asked for accounts of apparent premonitions, and received about 75 replies. Half of them were dreams and the rest waking hallucinations or a sense of foreboding. Mostly they came from the London area because of the newspaper's coverage. An exception was a dream of 'something black coming down over the school' experienced the penultimate night before the disaster by one of the children killed; at breakfast she told her mother about it who, understandably, paid no attention. Among the images reported were children, schoolhouses, Wales, coal and steep hills, some of which I give below. They demonstrate clearly that the foreknowledge was vague enough to be presented exceedingly generally, but for all that the event to which it related seems plain enough. Asterisks mark the experiences that were mentioned to another person before the news became known; the latency is the interval between the intimation and the event.

None of the accounts relating to Aberfan was written down before it happened. Of course there could have been embellishments after the event, especially in the instances which were not remarked on beforehand. Nevertheless, if any of the observers had been trying to stake fictitious claims to startling foreknowledge they would have found it easy enough to be more imaginatively comprehensive. (*Aberfan* ψ79.)

Table 2

Foreknowledge of the Aberfan Disaster

Images	Presentation Latency	Relevance
Terrifying dream of children trying to escape from a room; hundreds of horror-struck weeping people running to the same place	Dream 12 hours	Distress, children

Screaming children buried by an avalanche of coal in a mining village coal in a mining village	Dream 1 week	Distress, children, turbid flow, steepness, blackness
A school, screaming children, and 'a creeping black slimy substance'	Dream 2 weeks	Distress, school, children, blackness, turbid flow
The name Aberfan and desolate rows of houses	Dream 1½ days	Place, joylessness
Children in Welsh national dress going to heaven	Waking image 12 hours	Children, Welshness
100 children engulfed in black mud	Conviction 1 week	Children, blackness turbidity
Crowds on a hillside, mud everywhere, man holding lamp with child beside him	Dream 2 weeks	Turbidity, steepness, child
A descending mountain with flowing surface, rescued screaming child	Dream 1 day	Steepness, flow, distress, child
Frightening dream of children by a building below a black mountain, hundreds of black horses thunder down dragging hearses	Dream 2 weeks	Distress, children blackness, steepness, death

Mountain moving, black shale, children's screams	Dream 2 hours	Steepness, flow, distress, blackness, children
Screaming child in phone box; followed it to a house shrouded in black steam	Dream 6 hours	Distress, child, blackness
Vivid fantasy of the name Aberfan	Waking image	Place
Little Welsh girl saying "Aberredfan"	Waking image 3 weeks	Child, Welshness, Place
(By a resident of that town) Bognor disappearing under a mass of black pitch	Dream 1 month	Blackness, turbidity

A sceptical friend said to me: "So, in a population of several million what are the chances of several persons having a disaster dream on the same night (or several nights)? Quite high, I'd say." I like him well, but I have to say that this looks to me like generalising to accommodate a principled objection; the issue is not 'a disaster dream' but one concerning a black spoil heap slumping over a Welsh village and killing a lot of children. The chances of that particular event is very low. It has not happened before or since. Precognitions relating to Twin Towers were equally specific to that one event. Dr Sheldrake says he received more than fifty of them.

As it is, all the reports sound just what they purport to be – visions and dreams. All appeared to be related to the coming event but there were few hard facts and no homogeneity, although most feature children and distress in the dreamer. Striking apparent foreknowledge of some aspects of that single horrendous event seems to have been presented to different people by a wide variety of images that were almost incidental personal associations. Although each experience is light on detail collectively they suggest Aberfan very strongly.

* * *

Quantity of Information Conveyed

Usually very little apparent foreknowledge seeps through to us. Unpredictable events like horse races are illustrative just because they are simple. On the turf in real life largely uninformed guesses produce the winning animal for huge numbers of punters, and some will get the first two or three in order whenever enough people are betting. Correctly *dreaming* results, as opposed to guessing them, is of course another matter. It cannot be common if bookmakers are to flourish as they do, but such dreams are not equally improbable for everyone. A racing dream might not be especially noteworthy in an enthusiast who knows the names of the runners, but very extraordinary indeed in someone who does not. That is why this table includes the dreamer's familiarity with racing.

Foreknowing race results fascinates for obvious reasons, and several are on record. To add to Godley's experiences here are half a dozen in chronological order. *(Other horse races ψ80)*

Table 3

	Nature of the Precognition	Knowledge of Racing	Race Result	Source Remarks
1	A man dreamed of seeing the results board at Epsom showing the first three horses in the 1889 Derby, with *Blink Bonny* the winner	He had already bet on *Blink Bonny* at 10:1; confirming friends would not back such outsiders	Hours later *Blink Bonny* won; all 3 finished in the dreamed order	SPR; the second horse came in at 66-1 and the third was also an 'extreme outsider'

149

2	A woman was dreaming of visiting Lincoln cathedral when a man said 'Outram has won the big race'	She knew nothing of racing;	A few months later Outram won the 1914 Lincoln	Reported by Lyttelton in 1937; friends confirmed the dream report
3	Asked 'Who will win the Derby' woman's unthinking reply 'first 3 letters are P,H,Y, in any order'	No interest in racing;	Hyperion won the 1933 Derby a few weeks later	Lyttelton; confirmed by husband who had asked the question
4	An 80-year old Quaker dreamed of hearing the Derby commentary on the radio, including the first four horses in order	He disapproved of gambling and had no interest at all; he told two friends of the first two horses before the race	Also 1933 Derby, run hours later; all four horses were placed as dreamed	Saltmarsh; the dreamer forgot the 3rd and 4th horses until he listened to the broadcast out of curiosity
5	Over Christmas 1957 Godley dreamed the National winner as What Man third favourite at 25:1; later saw Mr What in a list of runners @ 66:1	No action due to long odds; on National day he saw Mr What now 3rd favourite @ 25:1; bet £15 to win, then £10 more	Mr What won The Grand National in 1958 starting at 18:1	Lord Kilbracken (Godley) in October 2000; Mr What's long odds explained by strong joint favourites

Table three contains a selection from about twenty reports I have seen. Racing precognitions suggest a limit to the quantity of information that is usually transmissible. The winning horse is always identified, though sometimes barely. In about half the cases where the dreamer does not know the runners beforehand the name is received only in part; sometimes the name can be garbled and sometimes only the jockey's colours are known. Two examples are much fuller; the 1889 Derby gave the first three and the Quaker got the first four. Much the most striking thing is the correctly disclosed names that were unknown to the dreamers beforehand – *Tuberose*, *Mr What*, the Quaker's horses and *Outram*, plus *The Oaks* that features in Chapter 16. All in all it is safe to say of race precognitions that the norm is very little information – all one needs on the turf – but there are occasional spectacular exceptions.

Aberfan is a more complicated event but it still shows similar paucity of material. Most of that foreknowledge comprises some or all of four elements only – steep terrain, blackness, children, tragedy. Out of steepness comes mountains and hillsides and avalanches; blackness is undefined or rendered as coal, pitch, black slime, black horses, black steam, black mud or shale. A school appears in one instance, a child or children in most, and distress in nearly all. There are some extensions to that limited content, notably Welshness and the name Aberfan (though the three instances of the latter were none of them mentioned beforehand). Two cases were largely symbolic, the first of children in Welsh national dress going to heaven, the second of black horses dragging hearses downhill, vivid imagery by dreaming assonance. It is curious that the salient point, the spoil-heap, never featured as such.

Sometimes so little information is passed that we miss it altogether. In wet mist one day I needed to wipe my brow; both hands being full with shopping bags I lifted my right to use the back of that hand, turning my head past it from left to right. As I did so, I remembered making exactly that movement for that purpose in a dream the night before that had been set in a steamy sauna.

On one occasion Dunne was shooting alone on a friend's land whose boundaries he did not exactly know. He found himself being shouted at from a distance by two men with a furiously barking dog and, worried that he might be trespassing, he slipped through a gate before they could come up. His notes for the previous night included the words:

(*Hunted by dog ψ81*) 'hunted by two men and a dog'

– but even with this reminder he had no recollection of any dream. Instances like this (which mine in the previous paragraph echoes) led Dunne to believe that fleeting and very slight foreknowledge is more common than we suppose. He went close to saying that hiding behind the trivial content of most precognitions might lie the possibility that they are an unexceptional occurrence for all of us, they only seem rare because few are noteworthy and we forget everything that is not dramatic, spectacular, or comprehensive.

Whether it is received awake or asleep we may infer that precognitive information of any kind is usually scanty and vague by nature; comprehensiveness and accuracy are exceedingly rare.

* * *

Latency

Latency is the term that has been given to the interval between a precognition and its related event. Usually it is very short, typically a day or two, but there are occasional exceptions.

Dunne once dreamed that:

(*Flying Canoe ψ82*) he had invented a flying machine and he was sitting in it – a tiny open boat made of whitish material on a wooden frame; there was nothing to support the boat and it did not need to be steered.

Twenty years' later he piloted, on its maiden flight, an aircraft he had made whose controls were sited forward of everything in a structure of wood and canvas like an undecked canoe. The machine was 'inherently stable' and would fly itself. At take-off something unexpected happened which Dunne does not explain that denied him control of the plane's speed. He sat in it as it climbed by itself at full power knowing that the engine would cut out in three minutes or so and he could glide it down. When it did he found himself coasting in silence three hundred feet above a field of scampering cows.

In *Visions of the Future* there is this second long-latency example, which was also a recurring dream: A woman wrote to Hearne:

(*Parachutes* ψ83) From the age of about five years until I was about 30 I had a dream at least every two or three weeks about being in a field with an aircraft crashing or about to crash, and parachutes above me. I was always afraid and awoke crying. All my family and friends knew about this dream. About 10 years ago I was in this field (the field in the dream) when two jets collided above me. The parachutes were there and I was crying but it was because my dream had come true and I just couldn't believe it. No one was killed. I was in no way involved with anyone in the planes. I have never had the dream since.

In an instance set in Iowa in the next chapter the event did not materialize for twenty-nine years; the detail of the pub damaged by the crashing World War II bomber was reported in 1989 to have come to hand 'recently', presumably meaning about forty years.

Chapter 15

Intervention and personal futures; foreknowledge of post mortem events

Intervention to avoid foreknown events

On this central question Louisa Rhine makes the good point that recognising intervention is more than ordinarily difficult because if it succeeds then the foreknown event does not occur and the experience is indistinguishable from any other dream that does not materialize. A case in point is this:

> *(Caravan ψ84) A* man in Illinois who was given to precognitions dreamed of an unpleasant visit to his trailer home by two men who approached from the adjacent highway. His trailer was parked on the edge of a farm with its door facing the road. Because of his dream he decided to turn the trailer round next day, even though it was heavily laden and the electrical and plumbing services had to be re-connected.

No visit followed. Did he prevent it or would there never have been one anyway?

She was rigorous in her winnowing of proffered examples, insisting for the sake of objectivity that intervention implied preventing the event from happening at all, not just evading its consequences for the dreamer. Among nine that she allowed out of nearly two hundred were these three:

First, a woman with her five-year-old son went walking one day while staying with her sister. They lost their way and were following an unfamiliar path when the sister said:

> *(Waterfall ψ85)* Don't let Jeffrey run ahead like that, call him back; I just remembered something I dreamed last night... in

a place just like this... the path ended in a precipice and I was lying on the edge holding by the hands a child who had slipped over. It may mean nothing but call him back.

The mother did so, and soon after the path turned and ended abruptly at a sheer drop, a viewing point for a waterfall opposite. She wrote that had the child run on at the pace he had been going he would not have been able to pull up in time to avoid falling. In the second a mother dreamed:

(Pebbles ψ86) ... she was camping with friends in an idyllic spot when she decided to wash some clothes for her one-year-old son in a creek nearby. Having found a suitable place she realized that she had forgotten the soap and returned to her tent for it – the child meanwhile throwing handfuls of pebbles into the water; when she got back she found him lying face down in the water drowned.

She awoke in great distress. During the summer – she does not say how much later it was – she and her children were camping with friends when she decided to do some washing. The circumstances of her dream repeated themselves; the location was as she had dreamed and the child was wearing the same clothes, but as she turned to fetch the soap the child's action of throwing a handful of pebbles into the water reminded her of the dream. She abandoned the washing and returned much upset to her friends. In the third instance a Los Angeles street car driver dreamed that:

(Streetcar ψ87)... having waited for the lights to change at a particular recognized intersection he met one block farther on the number 5 car coming in the opposite direction. He waved to its driver, and as soon as it had passed a large truck painted bright red did an illegal turn across his tracks, each obscured from the other by the number 5, and there was a hideous crash that overturned his car. There were two men and a woman in the truck; both men died and the woman, who had intensely blue eyes, was screaming in pain and blaming him.

He was given that run when he reported for work that day. On his second trip he suddenly became nauseated while waiting at the lights at the same intersection, and felt worse on seeing the number 5 car approaching. On waving to its driver he recalled his dream, braked, and shut off the power. As he did so a red truck, but a small one with white writing on the side, shot across the path of his car, which would have hit it had it still been moving at speed. In the truck were two men and a woman, she giving him a startled look with the same blue eyes. He had to be relieved of his duty.

Louisa Rhine rejected the case of a young newly-wed who was anxious to return home but suddenly found herself too scared to board the train at the station. It crashed with loss of life further down the line but Rhine did not reckon it to be intervention because although the woman's actions avoided the consequences for herself they did not prevent the accident happening. Here are three more where the recipients avoided effects to themselves alone.

> (*Fireman* $\psi88$) A young man secured a job as a fireman that he had been hoping for. The night before he was due to start he dreamed three times of a newspaper report that a steam explosion had blown him out of a building, and that he had died later in hospital. He declined the job. About a week later the boiler did explode, his replacement was blown out of the building and died, and another man was scalded to death.

Rhine did not regard that one as intervention because the event was not prevented, nor did she accept the next three:

(Bank Manager's Suicide ψ89) At the start of the Depression in the 1930s a woman in Washington D.C. dreamed that a man in jail shot himself in the temple with a revolver. As she was telling it to her sceptical husband over breakfast she 'knew' that it was the cage at their bank where the president sat rather than a jail. She became so upset that her husband agreed to withdraw their modest savings, and having done so but before leaving the building the police arrived because the president had shot himself in the vault after squandering the depositors' money.

(Boat sinks ψ90) A couple with a two-year-old daughter had planned some excursion on a boat but after the gangway had been lifted the wife suddenly insisted it be replaced so she could get off. Her husband was annoyed but the boat was later hit by a freighter and sunk in fifteen minutes without any loss of life.

(Baby and mattress ψ91) A woman in New York dreamed that she heard a scream, turned around and saw her two-year old son falling from an open window. She then heard the siren of an ambulance in front of the house.

She awoke and checked that the child was alright. A couple of days' later she put the child's mattress out of the window to air it and pulled the window down to hold it. Busy in the next room she suddenly remembered her dream and ran into the child's room to find that he had pushed up the window and had climbed on to the sill. She grabbed him as he was about to fall; the mattress already lay on the ground below.

Precognition of personal futures?

Requiring that intervention prevent the future event from occurring altogether makes it an all-or-nothing affair and therefore very rare; but if intervention were a matter of degree then it could embrace the four immediately preceding instances, and many more as well.

Demanding total prevention would be a perfectly reasonable prerequisite on the basis that a precognition is of physical events; but if one supposes that precognitions are intimations of a personal future involvement or state of mind rather than foreknowledge of the event itself, that they are a pre-experiencing of one's own reactions and feelings towards what is to come, then intervention could be seen as action avoiding only the consequences of an event to oneself, and not as altering the world for everyone else. Were it so, intervention becomes a much less outrageous suggestion – it takes only an umbrella to protect expensive clothes in unsettled weather, not dispersing square miles of threatening clouds to stop it raining at all.

Almost all precognitions are confined to witnessing the related events or to hearing news of them. Where they are calamities there is an emotional response whether the event is witnessed or not, and extraordinary trivialities are noteworthy only for being personal experiences. It would be a great simplification to say that all precognitions are personal, but the available evidence is thin, as this shows:

Waking premonitions of danger are necessarily personal – Corbett's tiger was only hunting him, that snake was threatening only that woman's child. All the precognitions in this book can be read as personal information too – it is not proved but there are hints at it. When war broke out Goddard understood that the hangars at Drem had been repaired and that is what he had seen in 1935, but when he visited that airfield in 1964 he saw that it was not so, the hangars had been demolished and re-built with different roofs; similarly Dunne mis-read 4000 deaths in the article on Mont Pelée when in fact it was 40,000. Both those mistakes were personal to the dreamer and not world events at all. Precognition has to be the transfer of information, visualized or not, but there is no reason why that should exclude supposed or apprised information, even that which is wrong. In the 'blue cardigan' dream the foreknown event was a mental one that never had any physical existence whatever – the woman's precognition concerned a hallucinated image in her own mind that she alone could see and was most strictly personal to herself. This personalizing might bear on the provenance of the imagery.

To borrow from the physicists a concept without its context, we might perhaps put it that individuals are 'entangled' with their own futures – hence the personal relevance of precognitions.

There is more on this later.

Dreams of Post-Mortem Events

On the face of it, foreknowledge of events arising after the subject's own death ought to resolve the question whether precognitions are of a personal or common future, but I have found only four of them.

The first in chronological order is Abraham Lincoln's well-known dream about a fortnight before his death. It was recorded by Ward Hill Lamon, a friend and former law partner, who wrote 'I give it as nearly in his own words as I can from notes which I made immediately after the recital'. In the course of a discussion about dreams in the Old Testament Lincoln mentioned that he had had a disturbing one, and was induced to tell it:

(*Lincoln's assassination* ψ92) About ten days ago I retired very late... I soon began to dream. There seemed to be a death-like stillness about me. Then I suddenly heard subdued sobs as if a number of people were weeping. I thought I left my bed and wandered downstairs. There the stillness was broken by the same pitiful sobbing, but the mourners were invisible. I went from room to room; no living person was in sight, but the same mournful sounds of distress met me as I passed along...

I was puzzled and alarmed… I kept on until I arrived at the East Room which I entered. There I met with a sickening surprise. Before me was a catafalque on which rested a corpse wrapped in funeral vestments. Around it were stationed soldiers who were acting as guards; and there was a throng of people, some gazing mournfully on the corpse whose face was covered, others weeping pitifully.

"Who is dead in the White House?" I demanded of one of the soldiers.

"The President" was the answer, "he was killed by an assassin."

There came a loud burst of grief from the crowd, which woke me from my dream.

On 9th April, 1865 Lincoln was shot in the back of the head in a box at the theatre without ever knowing of his assailant's presence. The ball travelled to the right front of his brain but did not emerge. He died early next morning without regaining consciousness, the first president to be killed in office. His body was embalmed and laid to public view in the East Room throughout the third day after his death. The face was uncovered.

The next is also rather light on detail. It has often been published before, and much embellished in the process. The nearest I have got to the original source is identical and apparently verbatim copies of an item from *The Bendigo Independent* that was reprinted in two Melbourne newspapers, *The Age* for 9th of November 1870 and the *Melbourne Argus* for the 11th. This is how they read:

(*Craig-Melbourne Cup* ψ93)…it is said that the late Mr Walter Craig, of Ballarat, told some of his friends a short time before his death that he had dreamt that he saw a horse, ridden by a jockey wearing his well-known colours, but with crepe on his left sleeve, come in first in the Melbourne Cup. Now

Nimblefoot, Mr Craig's horse, won the Hotham Stakes on Saturday, and his jockey who wore crepe upon his sleeve in memory of the late Mr Walter Craig, was loudly cheered upon coming to scale.

Walter Craig ran a hotel in Ballarat in Victoria. In May he had failed to find a buyer for the unexceptional Nimblefoot, but he entered it for the Melbourne Cup, a major race run in November of each year, and sent it to train in Tasmania. Not long before he died he dreamed as the papers reported. He is said to have told several people of the dream next day, as well as his conclusion that it signified Nimblefoot winning the race but he not living to see it. In fact he died in August but the horse remained entered and ran under his colours. Come the day, the 11th, it won by a neck in a most exciting finish. The report in *The Age* was published two days before the race, that in the *Bendigo Independent* must have appeared on the day before that, the 8th; and *The Melbourne Argus* report came out on the morning of the race.

The third example is from Louisa Rhine, one of many spontaneous letters she received as her work at Duke University, became known. She does not seem to have followed up very many of them, though in this case it would have been difficult. The gist of what a woman in Iowa wrote was this:

(Iowa horse attack ψ94*)*…in 1918 when she was thirteen her grandmother was very upset one morning about a dream the night before. Her grandmother ordinarily paid no attention to dreams but this one was unusually clear and distinct. In it she had seen 'a black horse with a white face charge me, striking me down. It was in the orchard just north of our farmhouse.' We had no horses running loose, and none black with a white face.

The grandmother died in 1934, the daughter moved away and then returned in 1947. She took in livestock for summer grazing, among them two black mares, one with a white face. They belonged to a bachelor and were unused to women; they usually snorted and ran when she approached but she was never afraid of horses.

> One morning when I went into the orchard the white-faced one snorted, then began to circle me, wide-eyed. I yelled, swung my arm, and grabbed a broken branch. She squealed and rushed. I jerked off a rubber, threw it in her face... I whacked her with the branch which broke in several pieces but scared her long enough. I reached the gate somehow... and got back to the house. And then I remembered. Here in this kitchen Grandmother had sat, shaken, almost sick, describing her dream to us, and only a rubber and a broken bough had kept that long-ago dream from coming true. [Rubbers are galoshes].

Finally, a woman reported to J.B. Priestley that forty years earlier (the episode must date from the early 1920s) her fit 24-year-old brother told her of his dreaming:

> (*Footballer's funeral* ψ95) of a funeral cortege where no one took any notice of him. The sister's black wide-brimmed hat, the funeral turnout, and the route it took were particularized, and also red and white flowers. Three weeks later the man died from peritonitis following a football injury and, according to his sister's report, the proceedings were just as he had said. She wore that hat and the flowers were red and white because those were the football club's colours.

Lincoln's is a dramatic dream, but its few details are not particularly significant. The East Room is the largest of the public rooms in the White House and it was where the body of President Zachary Taylor had been laid out fifteen years earlier. Those images could have been Lincoln's own integration. The only certainly precognitive fact unknown to Lincoln was a president's death by assassination. Craig's precognition of his horse winning seems to imply access to a general future that is independent of individuals, though not very firmly. Possibly his experience was a contrived Freudian wish-fulfilment dream – of course Craig would want to see his horse win; *Nimblefoot*, however unimpressive to start with, did in fact improve enough to win the major race and one may suppose that Craig was receiving progress reports from Tasmania. Because of the thirty-year lapse between the Iowa dream and the event there is scope for chance to have brought it about, and ten or so years before writing of it does allow unwitting adjustments. Nevertheless, no attack by a horse on any member of the family had occurred in that orchard in thirty odd years, and the beast's colouring was uncommon. As always there is a lack of detail.

The footballer's dream of his own funeral carries little weight. Barring the flowers, nothing about it refers particularly to him or his injury and any other funerals he and his sister had attended in his community would have been similar.

The imagery attaching to Lincoln's lying in state could easily have been personal associations but assassination was precognitive information. For Craig; the information was the fact of the race win and a suggestion of death. The Iowa dream of the attack on the granddaughter is significant. It could have been imaged from the grandmother's own mind but the person, site, appearance of the horse, and its attacking, were all included. The latency of the experiences was ten days for Lincoln, three weeks for the footballer, about twelve for Craig, and thirty years for the Iowa grandmother. They all look like precognitions of after-death events but if the precognition is of a 'coming personal response' then ought there not to be in some sense something

163

that persists to do the responding?

Lincoln found his dream depressing, Craig was happy and the grandmother anxious. All their emotions were wholly appropriate to earth-bound living people and seem faintly absurd in the context of eternity. These three experiences do not establish any kind of 'afterlife' but as they appear now they undermine the attractive simplification that precognitions are of future emotions rather than of something external.

It is a great pity that there are not more of them.

Chapter 16

Reporting Reliability – deceptions, errors and chance

Deception and Mistakes

Precognition is so extraordinary that covering all the possibilities has to include a common sense review of lying. There are two broad categories of deliberate untruths as they relate to past events: complete fabrications without any foundation at all, and embellishment with fictitious details added afterwards. If there are compulsive liars about, they are thin on the ground. Ordinarily, everyone who lies, does so for a reason. People are generally truthful according to their lights unless they perceive an advantage to be gained by not being so. It is all a question of motive, and only financial gain or self-advancement fits the bill so far as precognition is concerned. No one obtains direct payment for precognitive dreaming but an income might accrue from writing compilations of cases. Serious authors of these collections report no more than two or three precognitive elements among the vagueness in each dream written up; their writings could be a lot more dramatic than they are, whence a presumption of truth when they are not. Similarly, the reporting of trials with negative results like Dunne's joint effort with the Society for Psychical Research is self-evidently honest. The records of that Society, which are not published for reward, contain hundreds of named and located dreamers and witnesses; if they are all fictions they represent vast unremunerated labours.

More plausible than deceit for money are fabrications by some crusading enthusiast for a pet notion, or a taste for notoriety in the wake of some astonishing event. But ordinarily, people who report precognitive experiences neither expect nor receive any benefit from them. Often it is the reverse; individuals with ordinary lives to lead face ridicule rather than fame from precognitions and conceal their experiences rather than draw public attention to themselves; Wing Commander Goddard told a few friends of his experience but thereafter kept quiet for fear of being grounded as medically unfit for flying. These people wish to impart to others who share their interest something that the world disdains; no one takes the slightest interest in the percipient personally.

Dreamers of precognitions are so many, so varied and so open that there must be honest people of good intent among them, but they are not necessarily free of inadvertent self-deception by genuine error. Such an error is only possible by supposing something to be a precognition when it is not – which outsiders can judge for themselves.

Misremembering is a greater obstacle. Few dreams are written down because few are recognisably precognitive until a related event happens later on, and that allows the unwitting embell-ishment of details in between. This possibility detracts from the evidential value of each report but does not extinguish it altogether because there must have been something to embellish. The inclusion of details that did not feature in the subsequent event also indicates honesty in those reports. They were not embellishments because they were wrong.

Misremembering hardly applies where the dreamed data can be calmly reviewed awake, in company, and at leisure. The recently conceived 'false memory syndrome' has no bearing on dream reports; it is confined to spurious recollections that have been fostered by others over a period of time and does not arise spontaneously (Hochman, 1994). There must occasionally have been prior notification of some occurrence that has been forgotten, but it cannot account for correspondences with anything unpredictable like a transport accident. Some think that 'déjà vu', a phenomenon seen as lapsed memories of earlier real experiences, provides a more acceptable explanation of dreaming precognition but it has no bearing when something remarkable happens for the first time.

Of the abbreviated reports in earlier chapters some personal ones, like Dunne's, read well in the original and there is no reason to doubt their veracity. Those logged by the Society for Psychical Research are supported by written statements from people besides the dreamer, and are as well verified as they could be. Others are not reproduced to so high a standard. Hearne's, no doubt for readability, are short and unsupported beyond the subject's sometimes saying that so and so had been told before the event, and the same is true of Louisa Rhine's reports.

There must be some reports of precognitions that are worthless but they cannot all be. Little weight can usually be attached to precognitive dreams that are only reported after the event; but dreams that are spoken of beforehand are easier to credit, and those that are written down or acted upon are undeniable so long as the subsequent event is reliably witnessed. Familiarity with this topic, as with any other, is chiefly what it takes to distinguish truth from falsehood, witting or not. Repeatedly discovering clear, well-written, and factually vouched for accounts by personally trustworthy people compels acceptance – not necessarily of foreknowledge, but of repeating real experiences that look like it. Not everyone is a charlatan or a simpleton.

And then there is the critic who judges the reports. So long as he holds precognition to be impossible then any apparent report of it must be mistaken, but just one single unarguable case dismisses that entire class of objections. Repeated instances that are not individually faulted modify perceptions and change the unpalatability of alternative explanations, so that some reports can become unexceptionable. It is a matter of familiarity, recognising some to be possible makes it easier to accept more. This is not seduction from the path of objective criticism but a matter of how great the anomaly is perceived to be. Most people disbelieved sightings of the 'Surrey Puma' a good many years ago but few would have troubled to doubt a ginger tom in Maple Avenue.

The value of any report is ultimately a judgement based upon some chosen premise. Scant, disjointed and confused information is the phenomenon we have to deal with; it would be easier if it were crisp and coherent but it is not.

Chance

If chance can account for correspondences between dreams and later events there is nothing to investigate. There are only four approaches to take:

One is to see whether some precognitive dreams have a quality that distinguishes them at the moment they are experienced and do not rely on the later event for confirmation, so making them an identifiable phenomenon that must have a cause.

Another would be finding instances that are intolerably improbable.

The third is to investigate non-randomness in the distribution of the effect in the population as a whole; if they stood scrutiny it would imply a cause and not an accident.

The fourth is to explain their mechanisms.

Ascribing a distinct quality is a subjective exercise that does not take us far on its own, improbability is hard to estimate, questioning distribution has its difficulties, and there is no explanation yet.

Recognisably precognitive in advance

Some people identify particular dreams as precognitive, or likely to be so, before their related event happens. Mrs Hellstrom's precognitive dreams and waking images in Chapter 10 were brightly coloured, Dunne thought that the factory fire dream should be recorded as soon as he awoke; Godley always made bets based on what he saw as peculiar dreams, and latterly wrote them down also. Those people could classify immediately on waking most of their dreams as normal but a few as of future import.

An illuminating test of that would be to find dreams which the dreamer considered precognitive at the time but did not in fact presage a real event – obviously a difficult matter. We have two from Godley, one of *Claro* winning the *Cambridgeshire* when it was unplaced and *The Pretence* and *Monk's Mistake* both winning when only one of them did, to set against six dreams whose content did materialize. His is good evidence because he has specifically said that apart from his seven racing dreams he had no precognitions in his whole life. It suggests a conviction that a dream is carrying content relating to the future is not wholly reliable – as we might expect – but for a few people conviction is fairly frequent. It is not weighty evidence.

High Improbability

If any conjunction of dream and event arises only by chance there cannot be the slightest hint of a connection between them, and the probability of dream-and-event as an entity would be at least the multiple of the odds against each one separately; but if the juxtaposition of dream and event does not arise by chance, if there is in fact some as-yet-undiscovered link between them that makes dream-and-event not a separate pair but an indissoluble entity, then the likelihood of any juxtaposition becomes much the same as the likelihood of the event alone and not particularly remarkable. In *The Oaks* case below, the elements of the event are so improbable in themselves that nothing else matters, but in more ordinary instances the odds against both the dream and what happened need to be reckoned separately.

Take Miss B and the three cows as illustrating the relation of a dream to an ordinary event. One could perhaps make quantifiable estimates of the frequency of cows in that lane by the station, of there being three animals, of that colour walking at that speed, the different formations they could adopt and of the length of time a farmer could hold out a stick at full stretch, thereby arriving at the odds of that sight being briefly observable at all through the comparatively narrow aperture of that gate in any fifteen second period. One might also estimate the likelihood of a rail passenger having the time and inclination to walk the length of the platform before the train arrived, so arriving at the odds of anyone other than the station staff seeing it, and then make a further allowance for it being Miss B who actually did so. The figure would run to hundreds of thousands to one, but so also would the odds against any of the innumerable other things that she saw that day, were each broken down into enough detail. None would have any significance unless it had featured in an earlier dream, but estimating the likelihood of her dreaming all those very details by chance the night before, including the gate in the foreground, cannot be done because there is no saying how many possible dreams exist in a pool from which that particular one could have arisen by chance. Nor does it help to say that the dream may have been occasioned by a similar waking experience a day or two beforehand because that gives a chain of three unexplained juxtapositions instead of only two.

A further difficulty attaching to events that become public knowledge, like Mont Pelée or Aberfan, is that from a large enough pool of people thousands might have a similarly precognitive dream concerning it. Everyone would come to hear of the event, so that every such dream would seem precognitive. ('Similarly precognitive' means conveying bodies of fact relating to the coming event whose scope is equally wide and significant in each case; it need not comprise the same material.) Even remotely useful computations are impossible for almost any precognition, but there is a class of exceptions involving previously unknown names that demonstrates extreme improbability. Mrs Schweitzer's dream of her son's death is one, others were *Outram* and the Quaker's dream of the first four horses in order, and here is one more extraordinarily improbable happening. It was only written down after the event but, assuming its veracity, it detracts from chance as its explanation by two quite exceptional details. It is part of Hearne's 1982 collection.

(*The Oaks* ψ96) A few years ago I had a mental picture of a race-course with tall trees either side. I saw the horses running and heard the names of the first and second places loud and clear over a speaker; 'Val's Girl' and 'Juliette Marney'. At that time I had no knowledge of horse racing so I asked around the family about trees either side of the course. The race turned out to be *The Oaks*, but the bookmakers had not yet got a list of the runners as the event was a long way off.

As the time neared they were listed. Of course the family started placing ante-post bets along with a few friends who got to hear of my premonition. I was quite nervous by the time of the race. However, the order of winning was reversed. *Juliette Marney* won and *Val's Girl* was second.

Imagine my surprise when my slip was taken in – the bookmaker paid me in error *Val's Girl* 1st, *Juliette Marney* 2nd. So for me at least the premonition was correct!

I have checked this to the extent of confirming *The Oaks* result; the English *Juliette Marney* and the American *Val's Girl* were respectively first and second in 1975; a French horse came third. As always it is a pity there is not more detail, particularly what is meant by 'The race turned out to be... '

A racing friend tells me that there were at that time about five thousand horses in training in the UK, Ireland, and a few foreign stables which entered animals for British races. There may only be a hundred or so that are qualified by age, sex, and calibre to be entered for a given race, but the names of registered thoroughbreds are never duplicated so that the pool from which to make an uninformed guess at a name is not limited to that event's potential entrants but to every possible name there is or ever could be – a number that must run into hundreds of thousands if not millions. Uninformed people dreaming on occasion the name of an undeclared runner is exceedingly improbable.

Concentration upon Individuals

The question of distribution still has a contribution to make to the debate about chance because instances of very great improbability are rare. Most people who report precognition report numerous instances of it; it is quite common for some but utterly unknown to most. Now, if the cause of anything is chance alone its distribution throughout the population *must* be even and random, but concentration on a few individuals is not random at all. If this apparent non-randomness stands scrutiny it would prove some sort of specific cause or identifiable predisposition.

Any approach via distribution depends firstly on precognitive dreams being a heterogeneous class of comparable things. Since it is possible to talk about them collectively they do comprise a class of a sort. And next, they must be clustered on individuals – we can see Dunne, Godley and Hellstrom as examples, and a number of other individuals claim to have them quite often – every year or two. Part of the difficulty is that a great deal of uncertainty attaches to evaluating how precognitive a dream is, but it is less difficult to do for horse races where every precognition can be judged by one very simple criterion – did the animal win?

(a) Godley's dreams. According to *The Independent* racing desk and the Horse Race Betting Levy Board about two and a half million

people bet regularly on horses in the United Kingdom; there are about fifteen meetings each week providing, at six races per meeting, ninety winners per week, between three and four thousand in a year. What are the odds against anyone dreaming a winner by chance any night? To illustrate a guess, say that they are ten million to one, meaning that those two-and-a-half million people would average between them one winner every four days. That is about eighty dreamed winners per annum, with a few extras for the Classics, including the National and the Derby on which about ten million people bet. If there were really even eighty racing supporters dreaming winners each year – more than eight thousand in a century – we and the bookmakers would know about it. We do not know so the odds must be longer than one in ten million, but let that assumption stand.

The issue is the fact of dreaming any winner just before a race. When Godley had his dreams after the Second World War there may have been fewer horses and meetings than there are now and a different number of punters, but if the odds of dreaming a race result one night are no higher than 1:10 million, then they would always be that. More dreamers means more dreams in the same proportion, more horses means nothing because the number of winners remains the same and the successes of most are never foreknown anyway.

Godley's dreams were spread over three years, say a thousand days. It is a thousand times more likely that any individual should dream a chance winner in a period that long than that he should do it on a single night. That is, on these estimates, one in ten thousand rather than the assumed one in ten million. But John Godley dreamed six – the odds against that multiplicity by chance would be ten thousand raised to the power of six – 10^{24} or a million billion billion to one. Whatever the actual odds against dreaming a winner by chance doing so six times is very unlikely indeed.

But that is only part of it. Godley had never experienced any such thing before this batch of dreams started, and then they stopped never to recur in his whole life, bar one more in 1957. It is inconceivable that there could be no reason for that.

(b) *Dr Hearne's Reports.* By means of a newspaper article in the *Sunday Mirror* in January 1982, Doctor Keith Hearne invited accounts of apparent precognitions by dream or otherwise. Four hundred and eighty people replied, and of those who were sent a follow-up questionnaire, eighty-eight responded. Twenty-nine of them claimed to have had between ten and fifty precognitive dreams each and eighteen of them more than fifty each.

Fifty instances per person is a very high figure, but the nature of the survey does not allow great reliance to be placed on it. One could be misled by supposing all precognitions to be equally impressive when they are not. Unless each experience is published in detail – and they have not been – some must have been so trivial that their related events were hugely more likely than others, even certainties over a long enough period of time. All one can say is that if even two or three sets of reports stood scrutiny for ten or a dozen dreams then the effect is not evenly distributed in the population.

There are at least two difficulties with any collection of experiences. The first, which might lead to over-statement of the effect, is that the respondents are a self-selected sample; and the second is that a great many people are disinclined to give better particulars. The Society for Psychical Research did a study in the late 1950s where only seven per cent of those who had volunteered an account responded to follow-up letters (Green, 1960); and Hearne's response rate was twenty-seven per cent. Given this reluctance, the frequency of precognitive dreams may be higher than anyone has supposed in the English-speaking population but still nowhere near enough to make them common.

Without good records one cannot in most cases judge whether a dream is genuinely precognitive, but the fact that Dunne and

Hellstrom each had dozens of listed and particularized dreams establishes that at least in their cases the concentration on them was not random. Dunne mentions about forty of his own dreams in his book. Mrs Hellstrom kept a dream diary, which I have not seen, that contains about sixty cases.

The oddity of these concentrations on individuals, for what the figuring is worth, is that chance can only explain a random concentration on individuals if apparent foreknowledge is all but universal. Either precognition, though uniquely inexplicable, is so widespread in the population that it must be a real phenomenon of some sort, or it is so rare that the occasional concentrations on individuals must have a specific cause and therefore be real also. Concentration could demonstrate non-random distribution, but knowing certainly the numbers who never have precognitive dreams, rather than assuming it, would enhance the evidential value of their clustering upon a few individuals.

Concentration on Women

Women report precognition more often than men do. Two thirds of Dunne's student subjects were women. In Barker's paper on the Aberfan disaster quoted in Chapter 14, four claims of foreknowledge came from men, the remaining thirty-one from women, a ratio of nearly eight to one. Of those who responded to Hearne's questionnaire, reports from women outnumbered those from men by about ten to one. Overall, eighty-three per cent of reports of all kinds of abnormal experiences collected by Louisa Rhine came from women and two thirds of those experiences were precognitions. Priestley found women outnumbering men by three to one. Half of Hearne's respondents commented on family proneness, two thirds of those citing mothers or maternal grandmothers who were similarly disposed. He believed the female preponderance to be genuine because he could see no reason why it should be biased. Between 1981 and 1984 the

European Value System Study Group (EVSSG) interviewed more than eighteen thousand people in Europe and the United States on a wide range of topics including their 'psychic experiences'. The topics under that heading were whether respondents had ever felt an uplifting spiritual life force, or felt that they had been in touch with a person who was either distant or had died, or that they had felt an awareness of contemporary events at a distance. None of the EVSSG's questions concerned precognition awake or asleep. The questions could be described as of a religious or 'other world' nature and they were wholly free of any idea of a temporal anomaly. In owning to such episodes females led males by five to four in Europe and by six to five in the United States, so that any predisposition among women was not great, and far less than the collections of precognitions by Rhine, Barker, Hearne and Priestley suggest. It may be important that the EVSSG respondents were properly randomized.

I have found no measure of the incidence of precognition – as opposed to 'psychic experiences' – in the population at large. There may be one or two like that unstructured survey of American students, where it was eight per cent, but not broken down by sex.

But if the greater number of female reporters does not indicate uneven distribution of precognition then men must experience just as many equally impressive instances of it, the only difference being that they do not report them. Why? The experiences are either precognitions or they are the most extraordinary co-incidences, why should anyone conceal the latter?

A case can be made that human thinking processes lie along a spectrum from logic and analysis to intuition and synthesis and that, although any individual may appear anywhere along it, women tend to congregate at the intuitive end of that spectrum and men at the logical end. If that stood scrutiny one might say that men are more uncomfortable with apparently illogical ideas than women, and are prone to pre-judge them in the light of what they perceive as a rational understanding of the world. One has to allow that they answered freely to the EVSSG because they were not assumed to be adopting a viewpoint or belief. If this were what is going on, the supposed unease of a logical mind would only arise if the experience appeared to be a precognition rather than a co-incidence – the one being freely reportable, the other not. Non-random reporting would thus be indicating either that precognitions really are concentrated upon women and therefore have a cause, or that they are equally distributed but men tend to perceive them as a class of distinct and peculiarly inadmissible phenomena to which they respond differently. Either way they are distinguishable existing entities.

The case against chance is this:

Some individuals can identify precognitive dreams by something other than the later materialisation of their content, indicating a distinctive quality to some of them.

There are a few precognitions of utterly unknowable and indisputable facts – like those horse names.

Godley's multiple dreams on a single topic by a man who never had foreknowledge of anything else refute chance by non-randomness; one man is a small sample but he establishes a possibility in principle.

Such indications as there are suggest extreme non-randomness in the population as a whole; no study has been made of the incidence among supposed non-reporters anywhere to compare with individuals like Dunne and Hellstrom, but the concentration on those two people is undeniable. So also perhaps with some of Hearne's respondents.

The marked difference between men and women in their reporting rates for precognition suggests either non-randomness or once again some peculiar distinguishing feature.

The choice with apparent precognition is either that it is not randomly distributed and therefore is a reality with a cause or, to accommodate those rare instances of multiple experiences by chance, that it is very common indeed and grossly under-reported. There are no other positions to take. For all the imperfections of the data – by the difficulty of collecting it – there is nothing to commend chance as an explanation of precognition, even in the absence of any other.

Nothing to commend – that is the nub. Distribution cannot disprove chance, and however much common sense dictates that the horse names do so, exceedingly improbable is still not formally chance-free. But the only counter argument is that precognition is impossible and therefore any apparent instance *must* be chance whatever the odds. The keystone of that position is the word 'impossible'; no one may use it without first excluding every alternative by knowing every-thing there is to know. No such polymath lives, the cause of the common images in collective hallucinations is at least one thing that no one knows. To say of precognition 'it must be chance' is to reach for a comfortable orthodoxy, but there is no compulsion to be so timid.

Chapter 17

Evaluating Reported Precognition

Any objections to the reports of contemporary images in Part One – as opposed to the interpretations put on them – stand on weak grounds. Who is to say what images appear in another person's mind? The same holds for dreams, whence doubts about any apparent individual precognition recorded before the event can only stand on discrepancies between what was dreamed and what actually happened; but even that criticism is weak. Since dreams of past experiences do not match their related event very closely there is not the least reason to demand that any precognition should exactly correspond to the future either. What Dunne said of his agitated-horse dream – that it would have exhibited nothing in the least extraordinary after the event but was very remarkable before it – applies equally to all other instances. Dreams are dreams.

Examples of the forty-five per cent of reports concerning calamities given in Chapter 13 that Priestley received, and the R101, the aircraft collision in Tenerife, Lincoln's assassination (the first president to die that way), and the Twin Towers, were none of them predictable but each was emotionally charged, widely significant, and easily understood once it had happened. He found more striking and persuasive the forty-five per cent which concerned trivialities just because they were so utterly inconsequential – like the 33 eggs, the perching sparrow-hawk and Pirandello's bat, and from other sources the upside-down umbrella, the horse *Outram* and the Bishop's pig.

In judging any reports one must bear in mind these difficulties in observing them:

Even among the most receptive of subjects the episodes are

very rare.

Their duration is usually short.

The amount of information received is very small and may be masked by dreaming irrelevancies.

Spontaneity is central to them.

Physical recordings are impossible.

Experimental procedures demand repeatability, the very antithesis of unaccountable spontaneous phenomena; the law's 'balance of probabilities' is the best alternative approach.

Dreamers are much handicapped as observers but no one else can do it for them. The whole topic has to be considered within the constraints of what they report or not at all.

Even if one accepts the implication of non-physicality from Part One, the paradox of being able to alter a supposedly foreknown future is still an understandable reason why anyone might decline to countenance precognitions. Since we suppose that time flows in one direction only, and nothing can move backwards from the future to the present, then information that allows us to alter what is going to happen constitutes a stark anomaly.

Were we to find such a temporal anomaly to be a reality it could have far-reaching implications, but that is no reason not to look at it; quite the contrary, it costs nothing and something possibly related to it may be becoming evident in another field entirely.

Chapter 18

Experimental Derangements of Time

In 1927 Heisenberg made the seminal proposal that an individual particle's properties are inter-linked such that the better you can determine one the less sure you can be of others. This fundamental idea is called the Uncertainty Principle, but that is not the best translation of the German word he used. 'Indeterminancy' would be better, (or even 'indeterminability' that Michael Frayn has in the title of his play) but it is too late to go back now.

Einstein was unhappy with the whole idea of probabilities in particle physics. There used to be biennial invitation-only physics conferences bearing the name of Solvay who was the philanthropist who funded them. I have read that at the 1927 Solvay conference in Belgium Einstein proposed at breakfast one day a thought experiment that would disprove the newly formulated Uncertainty Principle by allowing two complementary properties of a single particle to be known together; Bohr and Heisenberg's contingent found the theoretical flaw in it by the evening, and next day Einstein produced another one. Whether it was all as immediate as that I am not sure, but the process continued afterwards by post and in time Einstein found one that could not be faulted. It was based on the state known as 'entanglement' in which pairs of particles have mutually exclusive reciprocal properties but any change to the state of one particle is immediately reflected in the other without any connection between them no matter how great their separation may be. In some circumstances it is possible to generate a pair of such particles. Measuring a particular property of one of them – which alters that particle slightly and excludes the possibility of knowing anything else about its earlier state – tells you that the still undisturbed particle has the reciprocal value of that property – its opposite or alternative. If the undisturbed particle then had a different property measured one would have determined both the properties of each particle at the time of the first measurement.

The theoretical importance of the matter was that Bohr and Heisenberg's position implied that until measurement each particle had both mutually exclusive properties simultaneously – that is each one of the pair being in both places and carrying all possible alternative properties. Einstein's view was less startling, allowing the properties to be pre-assigned but their individual allocations as yet unknown. In 1964 John Bell from Northern Ireland devised a thought experiment that could determine which it was from three classes of observations on paired photons being combined in such a way that the sum of two would be unequal to the third if the particles were adopting their values only on measurement. The exercise was entirely theoretical and the measurements impossible in practice, but as soon as 'Bell's Inequality' was published someone spotted a variation involving four classes of observations that was actually feasible. The race was on to try it.

In 1982 Alain Aspect at the University of Paris South was awarded his Ph.D. for performing that experiment, and it was repeated at other centres afterwards. All showed that Einstein's gut feeling was wrong – rather than being innate the values for each are only adopted on measurement. Before that, each member of the pair has both mutually exclusive properties 'superposed' on it at the same time, a paradox that is inescapable. Particles in such pairings are said to be 'entangled'. Slight re-arrangements to the same equipment and very fast switching also proved that the particles react to measurements on their twin in less time than light could travel between them. The information is transferred at what were called 'super-luminary' velocities and position is one of the properties that can be superposed too so that each particle might be in both of two places at once until a measurement determined which should be where – for which the term was 'de-localized' (Aspect, 1982). Although Aspect's conclusions were widely accepted there was an unavoidable statistical element to his method that occasioned a little unease. In 1989 the Austrian physicist Anton Zeilinger measured the polarisations of three pairs of entangled photons to establish without any doubt that the state of superposition is real.

The ability to generate and detect individual entangled photons brought a range of experimental surprises. A very striking one is known as 'delayed choice'. There are a number of configurations of apparatus to demonstrate it, of which what follows is a generalized case, taken from a teaching aid for a graduate physics course at Stony Brook University, New York, based on a paper by S.P. Walborn and others (Walborn, 2002).

A single photon can be made to divide into an entangled pair. Photon A of the pair travels the upper path in the diagram to a device that imposes a particular polarisation on it, call it x or y, which a detector registers; photon B goes via the lower path to a double slit. If A's polarisation is x then without measuring it B's

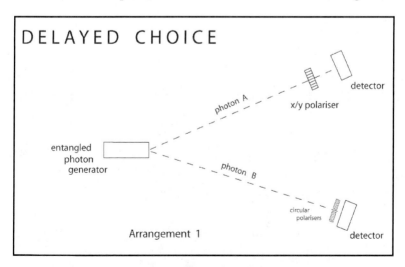

Figure 4

must be its opposite – *y* – because A and B are entangled. There is another kind of polarizer, circular polarizers, whose outputs *l* and *r* show a left-handed or right-handed twist, which direction depending whether the input polarisation was *x* or *y*. One of these is placed in front of one slit, its fellow in front of the other.

With this arrangement in place polarising A determines the circular polarity of photon B as it leaves each slit, meaning that the circular polarizers are carrying indisputable 'which path' information that annuls the interference; if A is not polarized the output of the circular polarizers tells you nothing because you do not know what their inputs were. Thus if A is polarized there should be no interference but if not then there should. That is what you see.

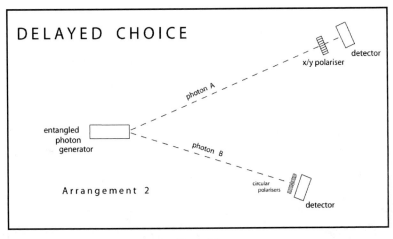

Figure 5

Now for the oddity. Increase the length of A's path to its polarizer so that it exceeds that of B's distance from the screen. As before, with A unpolarized there is no certainty and there is interference. But even if A *is* polarized B should reach its screen before that can happen, so that by common sense B's polarisation cannot be known in time and there will always be interference.

Not a bit of it. If A's polarizer is on and 'will' affect that photon when it gets there, then there 'will' be which-slit information and there 'will' be no interference. I use the inverted commas to highlight that we seem to be faced with a different kind of future to the one we are used to. Another way of putting it is that the effect – interference/no interference – is preceding its cause – polarisation/no polarisation; or that, for being entangled in the quantum sense, no reaction is complete until both photons have adopted fixed and certain properties by each separately making an irreversible mark on the physical world. Or yet again, that time does not flow within the apparatus, there is no sequence to the steps of the process, and the only 'event' is the completed whole – while for observers outside the apparatus time continues to run normally with its inherent sequencing of events.

There is no theoretical limit to the separation between entangled particles; a change to one is always reflected instantly in the other no matter how far away it is. As a pair they appear to be free of spatial constraints and the reversal of cause and effect implicit in delayed choice suggests they are free of temporal constraints as well.

Photon B's passage through the two slits and producing or not producing an interference pattern indicates the same spatial anomaly as in Chapter Nine – the photon seems to be nowhere – but in delayed choice there is a temporal anomaly also. There is no suggestion that these observations provide any answers to the questions posed by the psychological phenomena in Part I – what they do is expand the range of what is to be explained.

It may seem a little far-fetched to talk of another domain when all this is happening on a real laboratory bench in this one, but the concept of such a world can illuminate shared hallucinations too, in which the same imagery arises at the same moment in different heads right there. In both situations the cause of the phenomenon in front of us cannot be explained in the terms we use to describe the workings of the world. Whether you regard the causes of the observed effects as influences from another domain or something entirely surprising about this one is little more than personal preference. On the evidence space/time does not describe everything.

* * *

There is a great deal more to quantum mechanics than this, and both the illustrations here and in Chapter 9 are considerable simplifications. Researchers sometimes take years to verify and assimilate their discoveries, and the rest of us may only begin to comprehend them years later. We face the conceptual difficulty noted by Kant when he held that realities external to ourselves are as they are because of the way we observe them to be – that they can only conform to what our intellect comprehends. That need not preclude by degrees expanding or extending that comprehension in the face of entirely new discoveries, but it does mean that pending development of the necessary new concepts we are obliged to think by analogy. Initial thinking on the quantum world by physicists and then by the rest of us leads not at first to 'it is so' but instead to 'it is as if'.

The next section introduces a hypothesis that could accommodate both non-sensory psychological information and anomalous behaviour on a very small scale; and it examines the consequences of its being true.

Part III

Reconciling Differences

An argument to reconcile our understanding of the world with the need to engender, store and convey information on a future event, and then to allow the event to be changed afterwards.

Chapter 19

Fixity of the Future

Intervention, avoidance in the light of foreknowledge, seems at first glance to require more than one potential future and therefore more than one time. Such a concept leads to multiple times, popularly limited to three (as Dunne's were) because there are only three mutually right-angular spatial dimensions with which to draw analogies. And the authors who have dealt with such schemes believed an individual afterlife, arguing not only from observation but also from personal necessity. The idea of numerous times do not contribute to our understanding, but without them the so-called 'intervention paradox' – that a future avoided by foreknowledge cannot have been the future – seems to be an immovable objection. In fact it is not, it holds only if the future is fixed and unchangeable.

In some ways it certainly is. The sun will rise tomorrow at such a time precisely. The time and height of the astronomical tide at any particular port fifteen months hence can be known with certainty. On the other hand we have the common sense view of the Los Angeles street car collision - that the driver dreamed what could have happened had he not had the dream. However many details of his future that day may have been fixed, he had with respect to that crash at least the two free options of braking or not braking before he saw the van; they were not fixed at all. If the crash scenario he dreamed was really his future then it was obviously changeable in part.

The whole of our rational lives are predicated on changing and controlling the future –education, careers, economics, politics, medicine. All practical precautions share that same general aim – the Thames Barrage exists only to limit future flooding, and so indeed do all plans of any sort from military strategy to regulation of urban building. We believe unquestion-ingly that in public life anticipatory action is both possible and worthwhile, and on a personal level as well we all keep trying, with some success, to anticipate and amend. Such activity is really an attempt to alter the future by setting in train desirable processes in the present and letting them run. We may get it right and influence the future in the way we want, or we may not.

A changeable/fixed duality for the future is the nub of the matter, for future events fall into two distinguishable classes. There are those that are the unalterable consequential *outcomes* of processes running now, and there are *happenings* like the streetcar crash which seem to be random and need never come about at all. Return to the tide at that port. It is caused by the gravitational forces of the Earth, Sun, and Moon acting on water in that locality. They are acting now and changing in constant ways that will inevitably lead to the predictable result fifteen months hence. If there happened to be no tidal data for that area then thirty days' continuous observations would allow the presently acting tide-raising constituents to be determined in plenty of time to make a prediction for any time to come. Astronomy and cosmology deal mostly with events that are large-scale outcomes rather than happenings – for example such and such an asteroid will make a very close approach in 2215, and if we could measure well enough what is going on now we could know whether there will be an impact. An intermediate case is the departure of a train at 11.02 each morning, which is a matter of human intent; both the scheduling manager and the operating staff mean it to happen each day but sometimes it does not.

Quite different is the dreamed streetcar collision. Even though, had it occurred, it might have been possible to determine in the most minute detail how it occurred, it would have been avoided by either vehicle arriving at the contact point two seconds earlier or later. A passenger might have dropped her purse disembarking at an earlier stop, the driver's sneeze might have delayed his pulling away from the lights, a car in front of the van four blocks away might have been going marginally slower or faster – and so on almost without limit. Nothing *compels* us to suppose that life is certain on that scale, or to reject our intuition that such trivial details are matters of chance.

Although coming events may appear fixed in some respects and not in others, when the question of foreknowledge arises we adopt the simple belief that anything foreseen must be fixed. That is partly habituated faith in forecasts but it is also an attitude of mind inculcated by wonderful tales in myth and literature which show Fate's malign intentions to be inescapable despite knowing them in advance. If it seems uncomfortable to change that habit, view it in the light of Richard Feynman's remark on the quantum world – our difficulty is not that it is paradoxical but that our expectations were wrong. It is the same here; we know next to nothing about the future and are in no position to expect anything particular of it.

Were we to abandon in its entirety the notion that a fixed future is *necessary* to precognition, then the option of avoiding a foreseen future becomes a theoretical possibility, especially if all foreknowable futures relate to coming personal experiences. The future may be, as we ordinarily think, malleable in practice. If so it can be first one thing and then another; the purpose behind planning is to change the future from what it now is to something we would like better. What is more, it could alter after precognition of it, in the same way that seeing a blue front door to a cottage one day does not prevent its being green next time you visit that village. There are no grounds for pre-supposing that foreknowledge must fix events, a future might on a certain day show one thing happening on Wednesday week and two days' later something else. A changeable future necessarily implies changeable foreknowledge of it. A premonition that did not materialize need not have been wrong, especially if a consequence of knowing it had been deliberately avoiding the circumstances on which the foreseen events depended.

A minor apparent objection to precognition of visual scenes not previously encountered is that we cannot explain an image in our minds without sensing an optical stimulus to create it, that images of future scenes violate the order of cause and effect; but collective and successive waking hallucinations demonstrate exactly that process of non-ocular imaging.

These are the superficially salient aspects of the matter as they appear so far. There is a distinction between outcomes and events, and precognitions concern only the latter. The intervention paradox would vanish if the future were changeable; though nothing has been advanced to suggest that it is changeable, our experience of life gives not the least reason to suppose that it is not. The logical difficulties with intervention are entirely of our own making. Even though we endlessly try to influence the future in both public and private life, when we come to theorizing we insist as a premise that the future must be immutably fixed in one particular form. Were it changeable as we ordinarily suppose, the multiplicity of times that Dunne and others have proposed could be dispensed with and there would be no need to preserve them all in separate domains. Unless an immutable future is proved precognition need not be reckoned paradoxical.

We are left with that one starkly persistent question concerning any kind of foreknowledge – information backtracking to us through time from the future to the present.

Chapter 20

Time and Precognition

An essential property of the non-sensorily received information in crisis intimations relating to contemporary circumstances is its being conveyed from place to place, sometimes over great distances. In successive hallucinations like Miss Morton's ghost the effects repeat, so that there must be the storage of information for what can be long periods. Precognition also demands storage; the intimations in Chapter 9 relating to Aberfan, Tenerife, the R101 and the Twin Towers, arrived with different people at different times over days or weeks.

Seeing psychological information as a single whole is a small advance on Dunne's narrower position that dealt with precognition only. We can if we wish imagine ourselves set in three dimensions of space, one of time and, external to all of them, a non-physical domain of information. Its non-physicality could allow that domain of information to be contiguous with all parts of our ordinary physical world near and far, and coincident with all moments of it present and future. There would be no necessity for the information to be 'conveyed' in the ordinary sense of transported over distances or intervals, so long as it passes between that domain and the physical world we live in – which is just what it appears to do. Access to all space and to all time might conceivably permit non-physical information to be conveyed from one place to another contemporaneously in the present or from a coming moment to the present anywhere. Were it so, the temporal anomaly of backtracking through time would disappear. These are the difficulties with that idea.

A Venue for Future Events

It is hard to conceive any kind of information on an event existing unless the event has occurred, and an event is quintessentially a physical thing that demands a location and a moment in time. Every event is a re-arrangement of matter or energy at a place – the toast slips on to the floor, the switch is pressed and the light comes on. The re-arrangement cannot be erased and undone although it can be reversed by another event – the toast picked up or the light switched off.

Intimations of distant crises always concern contemporary physical events, and the same goes for shared hallucinations where the images are created at that moment. Each appearance of a ghost is a contemporary event also, even though the images seems to derive from the past. But information on an unrealized future event cannot possibly be generated in the present. It demands a world that is most clearly not this one. That concept is hard to grasp intuitively. Might a parallel universe be co-opted to assist?

Without exception every macroscopic entity that occupies the attention of physicists, material objects or fields alike, has a spatial location; the former have boundaries and dimensions, the latter attenuate with distance. Space is central to the discipline that is physics, without space there is no physics at all and that is why the key question is *where* even one *physical* parallel universe could be. It would have to be both huge and positioned, but the briefest glance around shows there is no room for it. There is no space for even a single separate material world with the required dimensions to occupy. The radius of our own world as we see it, fourteen billion light years or so, precludes any recognized influence from outside arriving here in our real world soon enough to effect a very short-lived process like one sub-atomic particle travelling a few metres through apparatus in a laboratory to 'collapse a probability wave form', i.e. select a particular possibility to materialize. To be workable such worlds must be supposed co-incident with this one – but that idea raises other difficulties. Distinct objects occupying the same single location that coincident universes imply is not a recognisable philosophical or physical concept. Uniqueness of position is central to either, and a unique identity is central to human experience; I do not feel myself to be one of who knows how many replicas.

There is another difficulty with parallel distinct but physical worlds of alternative futures. If one is to avoid an adamantine predestination that disallows intervention and free will, a physical future of events would demand a number of different futures, each housing one of the possible variations and each necessarily situated in its own separate world. A cook who scrambles an egg loses the option of boiling it, but might he leave intact another world where he does the other thing? That is all very well for a choice of two, but imagine three men among fifty soldiers being wounded in one of ten different ways capturing a hilltop. That would need nearly five hundred billion universes, 5×10^{11} of them and therefore as many worlds, should every trio of individual soldiers in that unit be allowed any one of those injuries. And that pertains to less than half a company advancing half a mile. What of the rest of the regiment? The brigade? The division? The army? And what of the whole war? Were another war to break out on a little planet in some immensely distant but just visible galaxy, is there a need for new universes undetectable by either of us, including our Earth, in order to cover all their eventualities too? We are running into an infinity – nothing untoward as a mathematical concept perhaps, but this is a solidly material one. It calls for limitless mass.

If one allows oneself to be beguiled by a parallel physical universe sited in 'other dimensions' we would need a unique set of three spatial dimensions for it, and multitudes of dimensions for a multiverse. In my view postulating any combination of co-incident or multiple *physical* universes does little to expand our comprehension.

Intervention and Conditional Events

The possibility of intervention disallows any kind of *physical* future because it implies that any events that may be foreknown are not fixities but only potential events at best. There is this to be said about them: picking up the child who threw the pebbles avoided a calamity on the lakeside, and all possibilities attaching to the boy falling over the cliff by the waterfall were annulled by the act of speaking. As a consequence of the streetcar driver's dream he acted to convert his vehicle's kinetic energy into heat in the brake shoes and wheels, and because he did so only two or three seconds before impact we can be sure he prevented a collision. In each case intervention by physical action avoided the *consequences* of a danger that had already arisen in outline, but the dangers themselves – the intention to leave the infant alone by the lake, the child running along that dangerous path, the close approach of the two mutually obscured vehicles – were not avoided. Only their outcomes changed. Bearing in mind that the imagery in frightening dreams may be alarmist, like Dunne's big black horse or the driver's big red van, one could say that death at the lakeside or falling over the cliff might have been subjective embellishments of that kind; if the first two of these precognitions concerned only the dangers and not their outcomes then intervention might have done no more than ameliorate its outcome. But the driver really did avert a physical impact by braking and the fireman really did owe his life to refusing that job. The wife's dream of a man shooting himself and her conviction that it was not in a jail but at their bank prompted her husband to close the family account. They were all material alternatives freely selected without compulsion by the people who were to be exposed to those dangers.

The materialising of any event in any present moment is confined to physical causes acting in an instant; a tree falls with the last blow of the axe. The supposition that an impending event inevitably results from present circumstances, that however distant in the future it may be its physical precursors are in being now and developing inexorably towards that end, is an unquestioned and venerable one, even though particular examples occasionally suggest very long chains of causation indeed. The attack by the horse in Iowa is one such, the parachutes dream is another and Dunne's canoe is a third – about 30, 25, and 20 years respectively. Though every world event has a preceding cause that is itself an event following from its cause and so on back, chains of events leading to a particular result lying undisturbed by chance intrusions for so very long a time must be exceedingly rare.

Perhaps that is why long latency is so uncommon. Observing that we can intervene means there is no compulsion for any particular possible future to come about just because it has been foreknown – an immaterial possibility can never be more than a conditional possibility. The common superstition that a foreknown future is unavoidable would be exactly the reverse of the case: if every aspect of the future exists only in non-physical form, and non-physical things are never fixed then the best we can say is that what we foreknow is a potential event only.

But allowing a foreknown future to change brings a secondary question in its train – how to account for the incompleteness of the precognition itself? Why is there no knowledge of the act of human intervention either, deliberate avoiding action does mark the physical world, it is a determining factor in the chain of causation leading to the circumstances that actually arise and it might be supposed as much part of the future as the event itself? Why is one link of that chain omitted and not the others? To put it another way, why is the 'wrong' future foreknown when inter-vention succeeds?

The chronology of dreamed precognitions leaves room for this idea: first a dream with precognitive elements arises, then it is recalled and is assigned to waking memory, where only then it leads to action. The particular synaptic connections expressing the memory trace that permits the intervention are realities that did not exist at the time of the dream. They only became a potential influence on the future on recollection afterwards, and that might never have happened, just as that imagined envelope and its contents might have been lost before it was opened. If a prerequisite for human intervention is a serviceable memory trace formed only on waking *after* the dream, then the chain of causation in being at the time of the dream had one more link added to it by foreknowledge – but only afterwards, and that might make the difference. It is unexceptionable that a future as dreamed before a waking memory of that dream is the outcome of the *then-existing* chain of causation which excludes the intervention sparked by the later memory that may never arise.

It is curious that in waking precognitions, like the Coptic Rose, the IRA bomb, President Kennedy, the pleasure boat that was rammed and even that member of the bomber crew, the subject matter did not allow any intervention by the person who received it. If last minute deliberate, or one can imagine purely accidental, physical action were to render a foreknown outcome impossible by a change to its necessary pre-conditions, then the future might change then and there, rather as physical action changes the state of an entangled particle at a distance. We all have a future of events, each one caused by its predecessors leading back from that future to the present moment. A future of some kind is certain; we are only ignorant of it because we do not know the links in its chain of causation. We could suppose that there is now and always was only a single approaching package of possibilities relevant to ourselves but its constituent details never enjoy certainty – only the physical present can confer that. If the subject matter of foreknowledge consists of conditional information it need only ever relate to a single version of the range of events that could constitute the outcome of processes running now – outcomes which could be altered by intrusions at any time before one of the possibilities materializes. Changing a foreknown future event allows us our cherished free will, but there remain those rare instances of events that seem to depend only upon receiving foreknowledge of them, cases where the precognition itself is one of their precursors.

One such is *The Oaks* horse race: assuming the woman's possible future was the bookmaker's mistaken pay out to her it could never have materialized unless she had bet; but even with the precognition she still might not have bet, like the woman who had the intimation of *Outram* winning the Lincoln. Her future with respect to *The Oaks* at the moment of her precognition was still only conditional. What she did was to put herself in the way of benefiting from what might come, just as does anyone who bets on foreknowledge; which is to say, she no more caused the error than punters cause their horse to win by betting on it; and the information presented to her was, for being wrong for everyone but herself, strictly personal. There are parallels with the parachutes dream. The chains of causation leading to those pilots colliding had nothing whatever to do with the dreamer; if the pilots had in fact been born when she had her first dream they were a long way from choosing flying as their careers. Her precognition was a personal experience as a spectator only; any calamity or breakdown to any aircraft involving any two people over any field like that any time anywhere would have done just as well. Her experience emphasizes that at least some precognitions relate to a coming personal state of mind and not to the objective world itself.

The crisis intimation 'Cousin Janie is dead' registered by that child in Scotland was caused by a real event but it can be thought of as tracking from South Africa via an immaterial dimension immediately adjacent to both places as *potential* information in that it only informs if it is received. It might have impinged on anyone with an interest in Janie's wellbeing but in fact only materialized as brain activity in the head of that one child. It is easily imagined that in most cases such potential information affects no one at all, and never actually informs. In the same way, only a few people receive foreknowledge of some public event like a transport calamity that may affect thousands. We can distinguish crisis intimations from precognitions by saying that crisis intimations comprise potential information on an actual event and foreknowledge comprises potential information on a potential event. Potential information and potential events would not be governed by any of the rules that relate to the physical universe.

An imagined Dimension of Information could embrace potentialities in some immaterial form. Perhaps the focus of our dreaming or inattentive mind drifts around our repertoire of recollections and may occasionally stray into the information dimension by something akin to Seligman and Yellen's process of adjacency; just as it makes nonsensical connections between past experiences it may also mince with them elements of potential futures that may never happen anyway.

The consequences of precognition are so profound that many people may be uneasy in accepting such a phenomenon solely on the basis of the tens of instances that have in the interest of brevity been advanced here to illustrate its commoner features. Keep in mind that there are very many more of them – the weight of the evidence is substantial.

A Hypothesis

Non-physically expressed information in a psychological context exists – the hallucinations, ghosts, precognitions and other experiences seen earlier in this book – it can be observed arriving and its conveyance is a necessary inference from that. The same information can be manifested on different occasions – which is to say it can be stored. Where we see it concerning yet-to-be realized events it is in some sense unconstrained by time. Its unexplained location and movement and temporal derangement are together what allows us to regard this psychologically manifested information as non-physical.

In the laboratory examples in Chapters 9 and 18 the perceived non-physicality is the behaviour of particles that is inescapably influenced by information that has no recognisable bearing on them. Aspects of that behaviour display spatial and temporal anomalies also, it is an effect without a cause.

Space and time are together as a single entity the foundation of physics and of material existence itself – everything that *is* has both location and duration. Entities without either one of them are non-physical.

The effects and nature of the two classes of inexplicable information in laboratory and psychological contexts are not the same, but a distinct domain that is locationless and timeless could satisfy both, and economy predisposes us to postulate only one unrecognized effect.

The hypothesis is that there exists an ambiance influencing aspects of the physical world that is outside time and space. It is not easy to understand, much less be proved, and it does not of itself explain the observed anomalies themselves, but it is a serviceable concept that has fundamental implications if it is true.

An Overview

Particles experimentally projected into a natural or contrived uncertainty behave in inexplicable ways, those in the ordinary world of certainties behave in ways we well understand. What exactly produces these different effects is still not fully explained, but the information that distinguishes certainty from uncertainty is physical – usually a signal from some instrument placed for the purpose. The effects which follow the arrival of that information operate only on a very small scale.

In psychological contexts strange information is producing measurable physical responses in human brains but the information itself is unaccountably non-physical and it relates to events on a human scale. The two sets of observations are not the same but both display spatial and temporal anomalies.

The scale of the spatial anomalies in physics is very small but in some of the psychological phenomena it is much greater, e.g. Miss Morton's ghost appearing all over the house and grounds and crisis information travelling thousands of miles. Similarly with time, in laboratories the effect is very immediate, but the specification for a ghostly image may be stored for years and precognitions seem occasionally to span decades.

Precognition is the most perplexing of all these anomalies. The present moment can be seen as the instant that a pre-existing but uncertain potential future collapses for physical reasons into the hard and certain reality of an immutable past in the world of events. The present can be thought of as the moment of phase change from uncertain to certain, rather akin to the change of state when a liquid freezes. Assuming that time cannot be reversing in some instances but not in others, that its flow is strictly unidirectional, then the receipt of even the most conditional of intimations of unrealized circumstances before they actually happen makes unexceptionable the proposition that time does not flow everywhere, that there must be aspects of reality outside it. Any kind of future which can give rise to non-physical information must exist in a domain or dimension where time is absent altogether, and whatever adjustments to the grand scheme of things that we might like to make to accommodate precognition must leave intact everything that physicists have ever discovered.

It follows that time has to be contained within the universe rather than the universe contained within time. Recapitulating:

There is no unavoidable principled objection to foreknowledge, particularly of an approaching personal state of mind; the whole case depends only upon the plentiful evidence for its happening.

The intervention paradox demands a conditionality which cannot exist in a state of coherent materiality – one event can be erased by another event but it cannot be undone.

A timeless and locationless non-physical aspect of the universe meets these requirements.

Were it so, time and space would be properties not of the whole universe but of only the physical part of it.

* * *

Where Now?

It will be hard going to take forward the hypothesis of an immaterial domain. The difficulties that have already been encountered in attempts to transfer psychological non-physical information in rigorous and repeatable conditions do not hold out much hope of progress along that line; attempts to induce precognitive dreams in selected groups have not worked well. Emotion is apparently an essential feature of all crisis cases and about half of all precognitions as well, so it will have to be accommodated somehow, but emotion is almost by definition beyond the purlieus of reason. Hypnosis looks a more promising lead towards a preliminary understanding of non-sensory information; those nineteenth-century transfers of tastes would be easy to verify, and if they stood up the same could be expected of smells, simple sounds, blocks of colour. What else might be transferable? Over what distances? Much might be achieved quite cheaply.

As for the spontaneous receipt of non-sensory information, recent examples attested as well as they were in the nineteenth century would be useful, as would more *post-mortem* precognitions. I have a feeling that close analysis of non-sensory imagery in every kind of experience would have more to tell us if enough reports – all of them – were to be examined. That little girl in the blue cardigan is nothing but a mystery as it stands, and the ghost cat that moved away with the daughter to her new house seems unusually odd. A deliberate large-scale survey like the European Value System Study Group to determine the prevalence of precognitions in a genuinely random group of people might be worth the effort.

Any impact of an immaterial domain on other fields could be illuminating. Over the fence so to speak, you would think that timelesnesss and locationlessness might bear on cosmology. And one can wonder whether non-physical information could contribute to the inheritance of complex behaviour in simple creatures.

The trail from the collective hallucination of that downhearted soldier in the foreword leads not to the notion that non-physical information exists – that was fairly evident early on – but to the entirely unexpected conclusion that there may be more to the universe than is confined to the domain of space-time. *Knowing* there is an 'other-world' aspect of the universe, one that has lain since the dawn of the scientific age outside the focus of rational human enquiry for its timeless and spaceless uncertainty would be a great advance on just believing it, and even the haziest idea of its nature would suggest new questions. The psychological oddities exemplified by the soldier can lead by an unexceptionable chain of thought to a totally unexpected hypothesis which is so far-reaching that the mere possibility of its being true makes it noteworthy. We are placed somewhat as nineteenth-century physicists were with radio waves, on the face of it there are unidentified and hitherto unnoticed but ever-present influences. Curiosity demands that we find out what they are, how they work, and whether they are useful; but they are an addition not an alteration, there is no reason to expect the least derangement of our philosophy of life or of science. Discovering such influences is surprising, astonishing even, but not perturbing. Whatever may emerge on further enquiry, the comprehension forged by the well-founded hard sciences will remain as secure as ever and we may advance from it with complete confidence.

Compared to the vast array of accepted fact and theory in all the sciences, non-physical information is only a most minor little wrinkle in the fabric of our understanding, but it is at odds with the rest of the pattern and we ought to know why. If information, no matter how conditional and in outline it may be, can exist on events before they happen then in some respects the universe differs from what we now suppose it to be.

That frog is twitching again.

Afterword on Religion

How might a non-physical domain relate to religion? Here is one view:

Richard Dawkins' book *The God Delusion* is a lucid and comprehensive non-believer's overview of Christianity, and in general of other religions too. He allows there are perfectly good reasons why a rationalist could adhere to the cultural values and customary practices of established churches but, likened to a sight-seeing tour, it is as if the reader was only boarding the bus at the second stop and missing a close-up of the genesis of it all. That is to say, there may be a better reason than the one he offers for the very striking fact that religion is a common feature of every population ever encountered, and so far as we can tell of every extinct one too. The reason is that people generally have always accepted in their everyday world influences that come from outside it. They did and still do heed religious leaders who purport to address these matters, which they see as immediately relevant to themselves in a way that quasars and muons are not. The ranges and behaviours of the spirits and deities which have over the ages been ascribed to this external domain are cultural artefacts of the civilisations concerned, but the venue for those beliefs is apparently enduring and universal.

Belief in personal post mortem survival is less soundly based. The analogy with the lost poem puts it simply; the work itself is not physical, but without any record it ceases to exist altogether on the author's death. Everyone can see that physical representation is essential to existence in the world but the entity represented need not be physical at all – an insight which may have combined with an immaterial domain in times past to spawn belief in an afterlife.

The modern variant of this domain is the parallel universe, an idea first proposed by Hugh Everett in 1957 to resolve a difficulty with Schrodinger's equation and gradually adopted more widely. For being a recent and secular concept it can be easily amended in the light of new discoveries, unlike the immutable 'other worlds' of religion. Each is meeting a need in its own context, the one intellectual, the other emotional.

Sources for Reports of
Psychological Experiences

No. (ψ) Ch.1	Subject Matter and Cited Works below	References to *PSPR, JSPR*
1	Robert Bowes	Bennett, p. 37
2	Mrs P., naval officer	*PSPR* 1889, vol. 6, p. 126
3	Mr and Mrs Barber	*PSPR* 1922, vol. 32, p. 172
4	Floating figure	Green & McCreery (1989) p. 163
Ch,2		
5	Miss Morton's ghost	Morton
6	Black cat	Green & McCreery (1989) p. 63
7	Grey-barred cat	*Ibid.* p. 61
8	Trondheim Nun	*Ibid.* p. 35
9	Seated Child	Personal communication
Ch.3		
10	Mr Young	Green & McCreery (1989) p. 35
11	Child burned	Rhine (1981) p. 47
12	Dead grandmother	*Ibid.* p. 63
13	Mother's heart attack	*Ibid.* p. 123
14	Jack slips at work	*Ibid.* p. 183
15	Injury in Alaska	Ibid. p. 213
16	Friend dying	Gurney *et al.* vol. 1, p.213
17	Cousin Janie dead	*Ibid.* p. 245
18	Sister ill in Yorkshire	*Ibid.* p. 245
19	Davie is drowned	*Ibid.* p. 246
20	Lord Brougham	Green & McCreery (1989) p .30; Gurney *et al.* vol. 1, p. 394
21	Airman in India	*PSPR* (1923) vol. 33, p.170
22	'The Major'	Green & McCreery (1989) p. 96

Ch. 4

Ch. 5

Ch. 6

Ch. 10

47	Coptic Rose	Hellstrom; MacKenzie (1974) pp. 32-46
48	Goddard airfield	Goddard; MacKenzie (1974) pp. 81-98
49	Bomber shot down	Hearne (1989) p. 52
50	JFK assassination	*Ibid.* pp. 33 and 34 (2 cases)
51	IRA bomb	*Ibid.* p. 13
52	Klara's dream	Personal communication

Ch. 11

53	Maury guillotined	Brook, p. 161
54	Water on actress	Dement & Wolpert
55	Children spill water	*Ibid.*
56	Walking on river	Green & McCreery (1994)

Ch. 13

57	Fashoda	Dunne, p .40
58	Mont Pelée	*Ibid.* p. 42
59	Factory Fire	*Ibid.* p. 46
60	Lady Q's uncle	Hadfield, p. 224; *PSPR* 1895, vol 11, p. 577
61	Escaped horse	Dunne, p. 48
62	R101 airship crash	Lyttleton, pp. 131-136
63	Typewriter repair	Rhine (1981) p. 219
64	Grenade injury	Priestley, p. 216
65	Sailing accident	Hearne (1989) p. 31
66	Tenerife aircrash	*Ibid* p. 41
67	Twin Towers	Sheldrake (2003)
68	Bishop's Pig	Myers (1895), p. 486
69	Umbrella upside down	Dunne, p. 101
70	3 Cows	*Ibid* p. 84
71	Cartwheel boat	Besterman, pp. 199-200
72	Sparrowhawk	Priestley, p. 220
73	Thirty-three eggs	*Ibid.* p. 218

| 74 | Pirandello's bat | *Ibid.* p, 227 |
| 75 | Racing dreams | Godley |

Ch. 14

76	Mrs Schweitzer	*JSPR* (1891-92) vol 5, p. 322
77	Girl in blue cardigan	Priestley, p. 208
78	Titanic	*JSPR* (1912) vol, 15, p. 264
79	Aberfan	Barker; MacKenzie (1968), p. 50.
80	Other horse races	Godley; Lyttleton; Saltmarsh
81	Two men and a dog	Dunne, p.82
82	Flying canoe	*Ibid.* p. 94
83	Parachutes	Hearne (1989) p. 57

Ch. 15

84	Caravan	Rhine (1989) p. 114
85	Waterfall	Rhine (1955) p. 25
86	Pebbles thrown by child	*Ibid.* p. 34
87	Streetcar	*Ibid.* p. 28
88	Fireman refuses job	*Ibid.* p.26
89	Bank manager's suicide	Rhine (1981) p. 176
90	Boat sinks	*Ibid.* p. 114
91	Baby on window sill	*Ibid.* p. 117
92	Lincoln's assassination	Stevens, pp. 183-187
93	Melbourne Cup	*The Age* and *The Argus*
94	Iowa horse attack	Rhine (1962) pp.35-36
95	Footballer's funeral	Priestley, p. 234
96	*The Oaks* horse race	Hearne (1989) p. 60

Abbreviations

PSPR, Proceedings of Society for Psychical Research

JSPR, Journal of the Society for Psychical Research

Works Cited

Alladin, A. (2008) *Hypnotherapy Explained,* Oxford: Radcliffe Medical.

Aspect, A., Dalibard J. and Roger, G. (1982) Experimental test of Bell's inequalities using time-varying analyzers, *Physical Review of Letters,* vol. 49 (25), pp. 1804-1807.

Barker, J, (1967). Premonitions of the Aberfan Disaster, *JSPR,* vol. 44 (734), pp. 169-181.

Bennett, E. (1939). *Apparitions and Haunted Houses* London: Faber.

Bernheim, H (1900). (Herter, C trans.) *Suggestive Therapeutics: a Treatise on the Nature and Uses of Hyptnotism,* New York: G. P. Putnam's Sons.

Besterman, T (1933). Report of an inquiry into precognitive dreams. *Proceedings of the Society of Psychic Research,* vol 41, pp. 186-204.

Billet, B., Alonso, J-M., Bobrinskoy, B, Oraison and Laurentin, R. (1976). *Vrai et Fausses Apparitions dans L'Eglise,* Paris: Éditions Belarmin/P. Lethielleux.

Bowyer, K. (1977) *Hypnosis for the Seriously Curious,* New York: Jason Aronson, Inc.

Brook, S. (1983). *The Oxford Book of Dreams,* Oxford University Press.

Corbett, J. (1943). *The Chowgarh Tiger* Chapter 3 in *Man-eaters of Kumaon,* Oxford Univeristy Press.

Dement, W. and Wolpert, E. (1958) The relation of eye movements, bodily motility and external stimuli to dream content, *Journal of Experimental Psychology,* vol. 55, pp. 543-553.

Dunne J.W. (1927) *An Experiment with Time,* London: Faber

Empson, J, (1993) *Sleep and Dreaming,* London: Jacob Harvester Wheatsheaf.

Freeman, W. (1980) Frequency analysis of olfactory system EEG in cat, rabbit, and rat, *Electroencephalography and Clinical Neurophysiology,* vol. 50, pp. 19-24

Goddard, V. (1966). *Light* (Journal of the College of Psychic Studies), vol. 86 (3465).

Godley, J. (1950) *Tell Me the Next One,* London: Gollancz.

Green, C. (1960) Report on enquiry into spontaneous cases. *Proceedings of the Society for Psychical Research,* 53, pp. 97-161.

Green C. and McCreery C. (1989). *Apparitions* Inst. for Psychical Research Oxford, pp 85-88.

Green C. and McCreery C. (1994). *Lucid Dreaming: the Paradox of consciousness during sleep.* London: Routledge.

Gurney E., Myers F., and Podmore F. (1889) *Phantasms of the Living* (vol. 1 and 2), London: Society for Psychical Research.

Hadfield, J. (1954) *Dreams and Nightmares* Baltimore: Penguin Books.

Hansel, C. (1980) *ESP & Parpsychology - A Critical Re-evaluation,* New York: Prometheus Books.

Hansel, C (1989) *The Search for Psychic Power – ESP and Parapsychology Revisited,*
New York: Prometheus Books

Hobson, J. A. (1988) *The Dreaming Brain,* New York: Basic Books Inc.

Hearne, K. (1984) A Survey of Reported Premonitions and of those who have them, *JSPR,* 52, pp. 261-270.

Hearne, K. (1989) *Visions of the Future,* London: Aquarian Press.

Hearne, K. (1990) *The Dream Machine,* London: Aquarian Press.

Hellestrom, E. *Correspondence concerning seven of her precognitions,* Cambridge University Library.

Lyttleton, E. (1937) *Some Cases of Prediction,* London: Bell.

MacKenzie, A. (1968) *Frontiers of the Unknown,* London: Arthur Barker Ltd.

MacKenzie, A. (1974) *Riddle of the Future,* London: Arthur Barker Ltd (Weidenfeld).

Middleton, L. (1989) *Prediction or Premonition? Coincidence be Damned,* Edinburgh: Pentland Press.

Morton, R. (1892) Record of a haunted house, *Proceedings of the Society for Psychical Research,* vol. 8: pp. 311-323.

Myers, F. (1895) The Subliminal Self, *Proceedings of the Society for Psychical Research,* vol. 11, pp. 334-593.

Orme, J. (1974) Precognition and time, *Journal of the Society for Psychical Research,* vol. 47, pp. 351-365.

Priestley, J. B. (1964) *Man and Time,* London: Aldus Books.

Radin, D. (1997) *The Conscious Universe: The Scientific Truth of Psychic Phenomena,* New York: Harper Edge.

Rhine, L. (1954) *The Reach of the Mind,* London: Penguin (first published 1948).

Rhine, L. (1955) Precognition and intervention *Journal of Parapsychology,* vol. 29, pp 1-33.

Rhine, L. (1962) *Hidden Channels of the Mind,* London: Gollancz.

Rhine, L. (1981) *The Invisible Picture,* McFarlaned & Co, Jefferson N.C.

de Saint-Denys, H. (1964) *Les Reves et les Moyens de les Diriger,* Paris: Claude Tchou (first published. 1867).

Saltmarsh, H. *Foreknowledge,* London: Bell.

Seligman, M. and Yellen, A. (1987) What is dreaming? *Behaviour Research and Therapy,* vol. 25 (1), pp. 1-24.

Sheldrake, R. (1999) *Dogs that Know when their Owners are Coming Home: and Other Unexplained Powers of Animals,* London: Hutchinson.

Sheldrake, R. (2003) *The Sense of Being Stared At,* London: Hutchinson.

Stevens, W. (1949) *Mystery of Dreams,* New York: Dodd Mead & Co.

Tonomura, A., Endo, J., Matsuda, T., Kawasaki, T. and Ezawa, H. (1989) Demonstration of single electron buildup of an interference pattern, *American Journal of Physics* vol. 57, pp. 117-120.

Tyrrell, G. (1943) *Apparitions,* London: Duckworth.

Walborn, S., Terra Cunha, M., Padua, S. and Monken, C. (2002) Double-slit quantum eraser, *Physical Review A,* vol. 65 (033618) pp. 1-6. Available as PDF: http://grad.physics.sunysb.edu/~amarch/Walborn.pdf

Newspapers from the National Newspaper Library, Cricklewood:

The Age and *The Argus,* Melbourne Australia, 9th and 11th Nov 1870 (Walter Craig).

The Catholic Herald, 17th December 1932 (Marian Apparitions).

The National Intelligencer Washington DC July 1850 and April 1865 (Deaths of Presidents Zachary Taylor and Abraham Lincoln).

Bibliography of Other Works

Books
Experiences

Behe, G. *On Board HMS Titanic: Memories of the Maiden Voyage,* The History Press, 2012.

Blackmore, S. *In Search of the Light,* New York: Prometheus Books, 1996.

Van de Castle, R. *Our Dreaming Mind,* Aquarian, 1994.

Connell, J. T. *Meetings with Mary,* London: Virgin, 1998.

Cornwell, J. *Powers of Darkness, Powers of Light,* Viking, 1991.

Graef, H. *Mary: a History of Doctrine and Devotion,* London: Sheed and Ward, 1965

Green, C. *Out-of-the-Body-Experiences,* Oxford: Institute of Psychophysical Research, 1968.

Guiley, R. E. *Harpers Encyclopaedia of Mystical and Paranormal Experiences,* San Francisco: Harpers, 1997.

Radin, D. *The Conscious Universe,* New York: Harper Edge, 1997.

Dreaming and Psychology

Bechtel, W and Abrahamsen, A. *Connectionism and the Mind,* Oxford: Blackwell, 2002.

Dement, W. *Some Must Watch While Others Sleep,* Stamford University Press, 1972.

Eysenck, H. J. *Sense and Nonsense in Psychology,* London: Penguin, 1957.

Flanders, S. (ed.) *Notes on the Dream Discourse Today,* Routledge, 1993.

Freeman, W. J. *How Brains Make Up Their Minds,* Weidenfeld, 1999.

Greene, B. *The Fabric of the Cosmos,* Vintage Books (Random House), 2003.

McKellar, P. *Experience and Behaviour,* London: Penguin, 1968.

Gross, R. and McIlveen, R. *Cognitive Psychology*, Hodder and Stoughton, 1997.

West, L. J. New York: *Hallucination*, Grune & Stratton, 1962.

Hamilton, M (ed.) *Fish's Clinical Psychopathology*, John Wright & Son, 1981 (reprinted).

Parish, E. *Hallucinations and Illusions*, Walter Scott, 1897.

Physics and Natural Sciences

Baggott, J. *The Meaning of Quantum Theory*, Oxford University Press, 1992.

Deutsch, D. *The Fabric of Reality*, Penguin, 1997.

Gribben, J. *In Search of Schrodinger's Cat*, John Wilwood House (Corgi Books), 1984.

Feynman, R. *QED*, Princeton University Press, 1995 (US); Penguin, 1990 (UK).

Lindley, D. *Uncertainty*, New York: Anchor Books, 2008.

Lindley, D. *Where Did the Weirdness Go?* Vintage Books, 1997.

Journals
Experiences

Cox, W. Precognition: an Analysis, II, *Journal of the American Society for Psychic Research* 1956, vol. 30, pp. 99-109.

Schouten, S. Analysis of spontaneous cases as reported in *Phantasms of the Living*. *European Journal of Parapsychology* 1979, vol. 2, pp. 408-455.

Stevenson, I. Precognition of Disasters, *Journal of the American Society for Psychic Research*, 1970, vol. 64, pp. 187-210.

Psychology, Physiology *etc.*

Berger, R., Olley, P. and Oswald I. The EEG, eye-movements and dreams of the blind. *Quarterly Journal of Experimental Psychology*, 1962, vol. 14, pp. 183-186.

Bressler, S. and Freeman, W. Analysis of olfactory systems EEG in cat rabbit and rat, *Electroencephalography and Clinical*

Neurophysiology, 1980, vol. 50, pp. 19-24.

Schaster, D. The Hypnagogic State: a Critical Review of the Literature. *Psychological Bulletin,* 1976, vol. 83, pp. 452-481.

Hochman, J. Recovered Memory and False Memory Syndrome. *Skeptic,* 1994, vol. 2, pp. 58-61.

Ohayon, M. Prevalence of hallucinations and their pathological associations in the general population. *Psychiatry Research* 2000, vol. 97, pp. 153-164.

Vedral, V. Living in a Quantum World. *Scientific American,* June 2011.

BOOKS

Iff Books is interested in ideas and reasoning. It publishes material on science, philosophy and law. Iff Books aims to work with authors and titles that augment our understanding of the human condition, society and civilisation, and the world or universe in which we live.